T0322413

SUNKEN LANDS

A JOURNEY THROUGH FLOODED KINGDOMS AND LOST WORLDS

GARETH E. REES

Elliott&Thompson

First published 2024 by
Elliott and Thompson Limited
2 John Street
London WC1N 2ES
www.eandtbooks.com

ISBN: 978-1-78396-769-8

Picture Credits:
All images © Gareth E. Rees, except for page 104: Used by permission of the
Edgar Cayce Foundation, Virginia Beach, VA.

Text Permissions:
Page v: Emmi Itäranta, *The City of Woven Streets*. Reprinted by
permission of HarperCollins Publishers Ltd © 2016 Emmi Itäranta;
page 45: 'Part II', *The Lowland Hundred*, The Lowland Hundred (Exotic
Pylon, 2014), reproduced with permission; page 142: 'The Tide', *Toll*,
Kemper Norton (Front & Follow, 2016), reproduced with permission.

9 8 7 6 5 4 3 2 1

A catalogue record for this book is available from
the British Library.

Typesetting: Marie Doherty
Printed by CPI Group (UK) Ltd, Croydon, CR0 4YY

MIX
Paper | Supporting
responsible forestry
FSC® C171272

To Kirsty

The world is ready to drown. The world is ready to rise.
On its surface walk creatures who have forgotten
their dreams, and only rarely do they remember
that their hours are brief
and their days are brittle, and there will not be
many chances at happiness.

Emmi Itäranta, *The City of Woven Streets*

CONTENTS

*Scan the QR code to listen to 'Songs from the Sunken Lands',
an immersive soundtrack to the book.*

See page 226 for further information.

1

CHILDREN OF THE FLOOD

One afternoon, a hunter named Nanabozho returned home from an arduous journey to find that his cousin was missing. He called his name, over and over, but heard no answer. His cousin's hut was empty. A half-eaten meal lay abandoned on the table. It was as if he'd vanished into thin air.

Nanabozho noticed a trail of churned earth and snapped twigs snaking through the forest. Immediately, he knew that his cousin had been kidnapped by his enemy, the Great Serpent.

The hunter followed the trail to a dark lake filled with evil spirits, including the Great Serpent himself. To lure the beast out of the lake, Nanabozho stopped the winds and asked the sun to beat down hard, hoping that the serpent would seek the coolness of the leafy, tree-lined shore. In readiness, Nanabozho transformed himself into a tree stump and waited.

As the lake warmed, the Great Serpent emerged, groggily, his scales flashing in the sunshine, and promptly fell asleep in the shade. Seizing his opportunity, Nanabozho shot an arrow into his heart. With a diabolical cry the wounded beast thrashed in the water, unleashing a great flood upon the land. As a tidal wave crashed through the villages and towns, drowning the terrified *people, the dying serpent rode high above on the foaming crest, his fiery eyes glaring.*

The flood waters continued to rise until only one mountain peak remained, where Nanabozho and a few other survivors took refuge. They built a raft to escape as the final piece of dry land vanished beneath the waves. Almost everything they knew was gone but for the birds that circled in the sky above them.

They drifted in despair for days and days until the waters began to recede and rugged peaks burst through the waves. Then they descended from the mountains to begin all over again.[1]

In 1984 an archaeologist was searching for submerged ruins in the Bay of Atlit on Israel's Carmel coast. Ten metres below the surface of the sparkling blue Mediterranean waters, Ehud Galili's team discovered seven megaliths on the seabed, clustered in a half-circle, tilted and eroded, but still standing after millennia under water, buffeted by the currents. Cup marks had been carefully carved into the stones, which surrounded the circular mouth of a freshwater well. Nearby, oval slabs were grooved with anthropomorphic symbols. Scattered across a 10-acre surrounding area were the walls of houses, paved plazas and graves full of bones. It was a lost world beneath the sea, untouched by human hands since it was flooded 9,000 years ago.

Gradually, Galili uncovered the remains of a village that was home to the farmers of a fertile plain, sown with wheat, grazed by domesticated animals. They crafted artefacts of bone, wood and stone. They buried their dead and carried out rituals at their sacred well,[2] sharing ancestral stories in the shadow of the megaliths until rising seas submerged their homeland, forcing them to migrate. But archaeologists also found a clue that something more sudden and catastrophic might have occurred: piles of fish, ready for sale or storage, lay abandoned. An Italian research team led by Maria Pareschi of the Italian National Institute of Geophysics and Volcanology suggested that a volcanic eruption of Mount Etna sent a tsunami pounding through the settlement, laying all to waste in a matter of hours. British marine archaeologist Dr Sean Kingsley even claimed that this could be evidence of the event that inspired the biblical tale of Noah's flood.

Atlit Yam is one of many sunken places around the globe that evoke wonder and speculation. Deep in the waters between Tunisia and Sicily, a 12-metre limestone monolith lies split in two, encrusted with barnacles, carved over 9,000 years ago, when that bank of seabed was an island. Off the eastern shore of the Croatian island of Korčula lies a 7,000-year-old road that was built by a lost maritime culture known as the Hvar. In the Arabian Sea stone pillars protrude at a location where the holy Hindu city of Dwarka, home to Lord Krishna, was believed to have sunk. Statues, goblets and sarcophagi are strewn on the sea floor of a bay west of the Nile, where the ancient Egyptian trading port of Thonis-Heracleion was guarded by the gigantic statue of Hapi, god of floods, taken down in a torrent of liquified soil when waters destroyed the city after an earthquake.

These are what remain of settlements from epochs when sea levels were hundreds of metres lower, their roads, walls and temples preserved in silt, along with tools, artworks and storage pits for meat and grains. These were centres of trade and agriculture, homes to gods, spirits and ancestors, abandoned to rising waters or destroyed by earthquakes, eruptions and tsunamis. Quickly, they were covered by sand, colonised by crustaceans and seaweed, stalled in time, while high above, on dry land, civilisations rose and fell. For millennia, the stone faces of forgotten deities peered at passing sharks through fronds of kelp. Lobsters scuttled over pottery fragments. Molluscs burrowed into tree trunks, their bark intact, sap still flowing in their capillaries. Jellyfish drifted over the bones of deer and wild cattle in peat beds imprinted with the feet of Stone Age children.

From time to time, these remnants would emerge at low tide after storms, snag in fishing nets, or rise to the surface. For instance, when the ruins of a Roman holiday resort were spotted in the Bay of Naples after an expanding subterranean magma chamber forced them up into the glittering shallows. Or when a night of savage storms revealed wooden posts in a circle around a Neolithic sky-burial platform on a windswept Norfolk beach. Or when a North Sea trawler dredged up a Mesolithic spear point in a clump of peat, revealing a forgotten land that once connected Britain to Europe before a tsunami flooded it 8,000 years ago.

Some of these places might have vanished from the historical record or sunk in a tumult of fire and flood long before humans had even a concept of history. But it doesn't necessarily mean that they were forgotten. Memories of settlements, hunting grounds and farmlands lost to rising waters, glacial floods and seismic cataclysms linger in folklore and cling to anomalies in the coastal topography. Flood stories form the root of mythologies across many cultures, going back to before ancient Sumerian times, and lie at the heart of all three major Abrahamic religions, hinting at traumatic experiences in our deep past, when the world became radically changed by runaway global warming.

Six hundred generations ago, our distant relatives experienced catastrophic flooding in a punishing series of climate shifts that began as the Ice Age came to an end sometime after 18,000 BCE, when the world's ice caps were at their furthest extent, known as the Last Glacial Maximum. First, there was a gradual warming period known as the Oldest Dryas, followed by the Bølling in which there was 300 years of super-accelerated heating, with melting glaciers, floods and sea levels rising by 16 metres. But just as the survivors adapted to the forests of birch and pine that flourished on once frozen tundra, the climate slammed into reverse. The world cooled for 600 years, sheets of ice surging back over the land. Another millennium of warming re-established the forests, but in 10,800 BCE Earth was plunged into temperatures as cold as the Last Glacial Maximum.

One theory behind the Younger Dryas freeze is that a North American proglacial lake, containing meltwater from the Laurentide Ice Sheet, was breached. A wall of water coursed through Washington State at 105 kilometres per hour, with aquatic tornadoes that scoured the rock like drills, carving the 100-kilometre-long Grand Coulee canyon. So much freshwater entered the sea that it altered the ocean's salinity and switched off the Gulf Stream, the warm Atlantic Ocean current. Another theory is that a comet hit the Greenland Ice Sheet, sending a cloud of dust into the atmosphere, unleashing the mega-flood and casting the world into 1,300 years of winter. Scientists have unearthed evidence of this impact in a 'black mat' of nanodiamonds, carbon spherules, iridium, charcoal and soot – all signs of a high-heat event – in over ninety sites in North America and Greenland, carbon dated from 10,800 BCE. Whatever the cause, the aftermath was devastating for life on Earth. Megafauna such as mammoths, giant ground sloths and sabre-toothed cats became extinct. Without a flow of warm Atlantic water moving northward, the ice sheets advanced and temperatures plummeted, rendering much of northern Europe and North America uninhabitable.

There may be traces of human experiences of this in some North American flood legends that exist today. The Pima from Arizona tell of an eagle that warned of a flood before a wall of water careered down the Gila valley, destroying all in its path.

The Choctaw tells of a darkness that fell over the earth, bringing unhappiness to the people. When their shaman spotted a glimmer of light in the north, they were full of joy, until they realised it was a mountain of water rolling towards them. Or there's the Ojibwa's tale of the Long-Tailed Heavenly Climbing Star, which burned all the trees and turned the world cold. The same tribe also have the legend of Nanabozho and the Great Serpent.

In 9600 BCE, the Gulf Stream was restored and the planet warmed again. Humans thrived as temperate deciduous forests returned, bringing nuts, fruits and edible plants. They settled on coasts and in river valleys, where there was an abundance of food and water. But with opportunity came peril. Global temperatures continued to soar and the seas continued to rise. Glaciers melted and proglacial lakes burst, unleashing floods that swamped low-lying Mesolithic plains. Peninsulas were reduced to archipelagos. Islands vanished. Land bridges disappeared. Forests turned to saltmarsh. As the colossal glaciers retreated, the land that had been depressed under their weight for so long began to rise. This phenomenon, known as isostatic rebound, unsettled Earth's crust, triggering earthquakes, volcanic eruptions, landslips and tsunamis that annihilated settlements, sank islands and reshaped coastlines in a matter of hours.

The popular perception of Stone Age people is that they were savage, club-wielding wanderers, hanging out in caves and grunting incoherently. But this is far from the case. They played instruments, made art, traversed seas, traded wares and studied the stars. These are the kind of people that experienced climate change in the early Holocene – folk who were not so different from us living today. They bore witness to frightening and perplexing Earth changes. Important places that were home to their gods and ancestors, monuments and hunting grounds, slipped beneath the waves. Communities were forced to migrate and begin again. It is likely that their stories about what happened were passed on down through the generations. In the north of Australia, for example, where the shoreline was drowned at the rate of 5 kilometres a year, Aboriginal peoples assimilated their experiences into their library of oral lore, the Dreaming. Many of its stories relate to flooding and describe land bridges that have long been submerged, including forests that stretched out to promontories that could only have existed between 9,960 and 13,310 years ago.[3]

There are over 2,000 known global flood myths, and these stories of catastrophic inundation can be found at the roots of many cultures and religions. For instance, when the Laurentide Ice Sheet collapsed 9,400 years ago, it dumped a massive volume of water into the oceans, raising the level of Mediterranean Sea. It breached a rock barrier in the Bosporus valley and entered the lake with a force 'two hundred times that of Niagara Falls,'[4] flooding 100,000 square kilometres of land over 300 days, creating the Black Sea. Some archaeologists believe this might have been the basis of oral traditions that influenced the Old Testament tale of Noah, gifted with foresight by God, who escapes with his wife and children in a boat laden with the male and female of every animal species. There is a similar story in ancient Greek legend where Zeus tells a man named Deucalion to build an ark before he floods the world, while Hindu texts from the sixth century BCE describe how the god Vishnu takes the form of a fish and tells the first man, Manu, to construct a boat before the deluge. It might have been one great flood that inspired all these myths, or a combination of cataclysms during the turbulent millennia of global warming between the Last Glacial Maximum and the end of the Bronze Age.

Natural disasters impacted not just hunter gatherers but civilised societies in built environments, such as the Mesopotamian city of Ur, constructed on a lowland swamp, flooded by the Euphrates in 3500 BCE. This might have influenced *The Epic of Gilgamesh*, an ancient Mesopotamian poem recorded on stone tablets in 2100 BCE, which tells of Utnapishtim, who was granted immortality after surviving a flood in a boat containing many species of animals. Even more calamitous was the volcanic eruption that tore apart the Greek island of Santorini circa 1600 BCE, home to a Minoan city of public roads, sewers and multi-storey houses, which some believe could be the source of the Atlantis legend told by the philosopher Plato. Afterwards, a giant wave crashed through the islands of the Aegean, heralding the end of the Minoan civilisation.

The stories told about these hugely traumatic, socially transformative events preserved them in the collective memory, where they became refined into myth, exaggerated and embellished over generations of retelling. Mutated traces of these ancient experiences entered the religious texts, philosophical works and folkloric literature that still resonate in our modern culture.

We are the children of the flood. All of us living today are descended from those who saw their lands drowned, civilisations crumble and populations scatter. Floods linger deep in our cultural memory. They ripple through songs, prayers and stories about a time of great disaster, when an old world died in violence, and a new one was born. I remember sitting in a circle of children on a musty rug in a schoolroom in 1979, singing 'The Wise Man Built His House Upon the Rock'. At the chorus line 'the rains came down', we mimicked falling raindrops with our waggling fingers and at the line 'the floods came up', we lifted our palms forcefully to the ceiling to represent the rising waters. Then we all clapped our hands as we sang 'The Animals Went in Two by Two'. I already knew well the story of Noah's Ark, and the lethal deluge sent by God. So I belted it out with gusto in the hope that such a disaster would never again befall humankind.

In my early childhood I was fascinated by sunken kingdoms and lost prehistoric worlds. I loved films like *The Land That Time Forgot*, *At the Earth's Core*, *20,000 Leagues Under the Sea* and *The Fabulous Journey to the Centre of the Earth*. They inspired the stories I began to write at the age of seven. My first, *The Travel to the Underwater Palace*, was scrawled in a lined notebook and began like this: 'An old man sat in an old rocking chair and told three sailors about an underwater palace. "I warn you, though," he said, "you will have to go through some dangerous lands."'

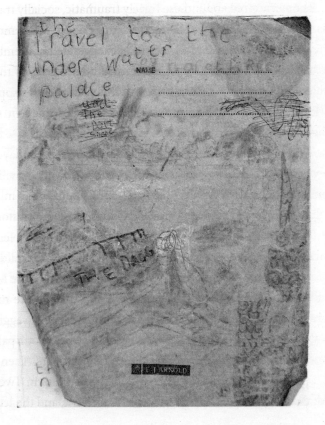

My story was about a crew of sailors, led by Captain Doom and Captain Boar, who adventure through islands populated by demons, dinosaurs and cavemen in pursuit of a fabled palace. I never made it clear whether it was built above water, then submerged, or constructed on the seabed. All I remember is that I fantasised about a city beneath the ocean. My cover art depicted the palace as an ornate tower, decorated with colourful fish scales, crested with a golden spire like a narwhal horn. When they finally spot the spire after a perilous journey, the sailors dance on the deck with excitement but Captain Boar stands at the bow, staring at it in silence. This was a strangely melancholic note in an otherwise mercilessly violent tale of derring-do. Perhaps it was my unconscious response to the eeriness of absence that characterises sunken places; that insatiable yearning to see the invisible.

I had recently moved to the English county of Derbyshire, in which the village of Derwent was flooded to make way for the Ladybower Reservoir, leaving only a protruding church spire to mark its watery grave. There are similar sights across the world: in Venezuela, where the spire of Potosi's church pokes above the water after the town was flooded for a dam; in Romania, where the village of Geamana was flooded to make way for a copper mine, leaving its church spire jutting from a toxic soup; and in northern Italy, where a fourteenth-century church tower looms over the waters of Lago di Resia. Either I had seen one of these images as a young child or I had intuitively used this iconic motif in *The Travel to the Underwater Palace*. Further evidence that I had unwittingly tuned into some kind of folkloric narrative current can be seen in my follow-up story, *The Secret Island*, which began like this: 'A boat sailed across the ocean, when an island rose up from the sea. The island had been under water for one hundred years.' This description was startlingly similar to the Celtish myth of Hy-Brasil, an enchanted ghost island in the Atlantic to the west of Ireland that was said to emerge for one day every seven years. I had never heard of that legend but perhaps my imagination had accessed a collective consciousness of flooded worlds, passed down through hundreds of generations since the Ice Age.

Decades after my schoolboy notebook novels, my relationship with sunken lands was rekindled when I moved into a house near the Lea Valley marshes in east London. Most of the marshland had been reclaimed for football pitches, reservoirs and industrial estates. But there were patches of waterlogged land, populated by herons and cormorants, untouched since the last Ice Age, offering a glimpse of an antediluvian world beneath the asphalt plateau of the city. The topography was haunted by fragments of the past, buried in the ooze. A dugout canoe. A rhino horn. A flint tool. Local legends told of phantom bears and crocodiles lurking in the undergrowth. It was as if the prehistoric past coexisted with the present. But it was the future that seemed to haunt the marshes most. The Lower Lea Valley is a floodplain, and it is only a matter of time before it is inundated. Scientific projections of sea-level rise put the area under water by 2050. As I wandered the marshes I imagined that one day a great wave might surge down the Thames and roll up the River Lea, submerging railway bridges, houses and warehouses, leaving only the tops of electricity pylons jutting from the surface. My first book, *Marshland*, described my strange experiences walking the Lower Lea Valley and ended with a speculative tale about this flood, where sea creatures slithered over drowned football fields, following long-lost tidal currents.

When I moved to Hastings on the south coast of England, I found myself in a nexus of sunken landscapes. To the east, the marshland of Romney lay hunkered behind grassy dykes and concrete walls. To the west, the Pevensey Levels, a tidal bay in

11

Roman times, was a flatland criss-crossed with water channels and ditches, prone to flooding, protected temporarily by 9 kilometres of shingle sea defences. I liked to sit on the cliff by the ruined Norman castle near my house and look over the English Channel, where Stone Age hunters once roamed a forest that now lies beneath the fish and the ferries. On the nearby beach of Pett Level, remnants of that forest were revealed at low tide, a chaos of algae-smeared stumps and severed branches, the grain of the ancient wood still visible. Out towards the shingle spit of Dungeness was the location of old Winchelsea, a medieval town that was destroyed by storm tides and forever covered by the sea. Behind it, sheep grazed on fields threatened with re-submergence.

This layered landscape told a story about cycles of change going back into prehistory, in which human beings had to cope with great loss and adapt to new circumstances. It inspired my second book, *The Stone Tide*, an autobiographical novel set on the crumbling East Sussex coast, about how the structures in our lives, which we perceive as being so solid and enduring, are actually transient and ever shifting. Afterwards, tales of floods continued to feature in my weird horror stories, like nightmares I couldn't shake off.

During the first year of the coronavirus pandemic, I spent time on the beach of Pett Level with my two young daughters. As the girls ran over the bones of the drowned Mesolithic forest, I felt a connection to that lost prehistoric world, as if we were part of a story ongoing since the Ice Age, when other fathers of other daughters faced up to the threats of their time as waters inundated their cherished homelands and their world became inexplicably transformed. I saw the ancient flood flow into a future flood in which this beach would one day return to the seabed and I wondered what would remain of this moment in time. What clues might someone find when the waters drew back again? What trace residues of our hopes and fears might exist in the story of this place? What might they learn of our joy and our suffering in a time of environmental breakdown and climate change? Because there was no escaping the reality that similar events were taking our own civilisation to the brink.

The following summer, I watched in helpless horror as the world flooded. Torrential rains in Europe swelled rivers until they burst their banks. Wild water rapids whirled down residential streets in the Netherlands, Germany and Belgium, hurling cars into houses and uprooting trees. The earth was turned into a filthy soup, flowing at terrifying speeds through villages and towns. 'Everything broken, swept away, it's a catastrophe,' lamented a seventy-six-year-old woman from the village of Schuld, as disorientated crayfish crawled down mucilaginous streets strewn with household appliances and rubber tyres.[5] In Erftstadt-Blessem, a sinkhole opened, sucking pavements and lamp posts into its deadly vortex. Cascades of gluey mud encased lawns, playgrounds and tennis courts. Gravestones poked from a sea of sludge in a decimated cemetery. Bridges on the River Ahr buckled in the torrent, cracked slabs of asphalt dripping with foliage and plastic bags. When the waters receded, cars lay upside-down in fields, trucks abandoned in drifts of splintered timber. 'My city looks like a battle has taken place,' said a resident of Rheinbach. Observing the grisly aftermath, an emergency chaplain from Bonn said, 'We live in a society that thinks it can control nature. And now people are feeling powerless against it. We have to be afraid of water and fire, like our ancestors 40,000 years ago. That's very difficult for people to understand.'[6]

That same summer, extreme rainfall triggered flash floods in Dharamshala, India, sweeping away buildings and cars. In China, floods in Henan province killed almost 400 people. A tunnel in Zhengzhou city filled with floodwater, trapping the terrified drivers inside it, while passengers in a metro train screamed as water seeped through the carriage doors. In the USA, Hurricane Ida's storm surge smashed through homes in the Mississippi Delta. Weeks later, a cyclone flooded Chesapeake Bay and water flowed through the streets of Washington DC, Baltimore and Annapolis. In October, a flood in Nepal killed over a hundred and displaced thousands in the Mahakali River basin. A gigantic plume of water vapour spread over Pacific skies and unleashed itself on Canada, submerging swathes of southern British Columbia. In the aftermath, the hulks of farm buildings loomed from floodwater lakes, bristling with telegraph poles, among the emergent humps of bridges and flyovers. Travellers were stranded in a town named Hope where all routes out were blocked by landslides. Meanwhile yet another high tide flooded Miami, Florida. 'The USA is SINKING!' proclaimed a YouTube video of cars moving through rising waters.

Floods were on the news every week, one flowing into another until all the world became a deluge, viewed on rolling social media newsfeeds from the sanctuary of my house on a hill. The video footage looked like the montages from those worst-case-scenario 'what if' documentaries I watched when I was younger: palm trees bent double in hurricane-force winds; wooden houses disintegrating into raging torrents; boats beached on roofs; waves sweeping through city streets. It was the stuff of nightmares. My worst fears about climate change were coming to bear in a way that seemed hyper-real, like a Hollywood disaster movie with exaggerated plots, accelerated timelines and disproportionate scales. It was suddenly easy to see how the experience of similar climate disasters in prehistory might give birth to epic myths and legends.

They used to call these 'freak weather events' but in the past two decades they have become the norm. Global warming has pushed the world's weather to new extremes. Hotter oceans and warmer air increase the frequency and intensity of storm events. Ferocious hurricanes are unleashing more rain than ever. As the oceans heat, the water expands and sea levels rise, making coastal towns and cities more vulnerable to storm surges. And worse is to come. Global temperatures are predicted to rise by more than 1.5°C above pre-industrial levels. Even if the industrialised nations were to take all the required action to reduce emissions, there is no way, say climate scientists, to stop the ice caps melting. The Greenland Ice Sheet has passed the point of no return. Half the glacial ice in Central Europe and North America will be gone in the next few decades, dramatically raising sea levels. Climate Central, which creates digital maps of projected submergences across the world, reveals that by 2050 the following cities could be under water: Amsterdam, Basra, New Orleans, Venice, Kolkata and Ho Chi Minh City. These aren't fringe communities but the densely populated nuclei of modern civilisation; repositories of cultural history, architecture and art, doomed to sink. And this isn't some speculative future catastrophe. It's happening now. In Jakarta, Indonesia, half-submerged houses have been abandoned to encroaching waters. Fish have colonised the waters in an abandoned shopping mall in Bangkok, a city under almost constant flood alert during the rainy season. In the low-lying areas of Mumbai, India, the underpasses and subways fill with water after heavy rains and waves crash over sea walls. In Alexandria, Egypt, winter storm surges pour into the cafes along the seafront where a concrete wall has been built as a last bastion against the inevitable.

Inspired by these troubling global events, I decided to embark on an exploration of sunken lands, past and present, real and mythical, to try to understand better our climate disaster and what it means when a civilisation faces radical transformation, or even annihilation, by unstoppable natural forces. For two years, I journeyed to drowned forests, shrinking wetlands, vanishing islands and sinking towns, beginning in the south of England, moving north-west to Wales, south-east to Italy and then west, out across the Atlantic to the United States of America. I followed the footsteps of the dead through sand, clay and silt, and placed my hands on rocks and trees touched by those who lived thousands of years before recorded history. I dived through shoals of fish to touch ancient ruins in Mediterranean waters. I engrossed myself in tales of lost worlds, destroyed by geological and climatic changes, from rising seas and tsunamis to earthquakes and volcanic eruptions. Some were the inventions of people trying to make sense of tree stumps that appeared on beaches at low tide, or anomalous post-glacial rock formations, presumed to be the ruins of legendary cities. Some were trace folk memories of real Stone Age, Bronze Age, Iron Age and Roman settlements, land bridges, islands and lowlands sunk by earthquakes and floods; transformed into legends like those of Ys, Lyonesse, Atlantis and the Lowland Hundred. In these tales, arrogant cultures dismiss warnings that they must mend their decadent ways and live harmoniously with the natural world, only for their walls to collapse and the seas to roll in. Similar narratives were being repeated in towns and cities flooding today; ongoing traumas from which shoots of future folklore might grow through the ages, just as children still sing of Noah's Ark.

As a writer I have long been fascinated by the way mythology and folklore evolve and adapt to new environments and cultures. The flood events occurring now may become legends in hundreds or even thousands of years, told by people who no longer remember our civilisation, but who intuitively understand the warnings encoded in the folklore. Perhaps such stories will help them avoid our terrible mistakes, much like those gnarly sculptures intended to warn future generations away from nuclear dumping sites, transmitting a universal message of horror that's likely to endure for epochs. This is why the tales we tell ourselves about what is happening in the world right now, and *why* it is happening, are so important.

To tell my version of the story, I visit an impoverished neighbourhood of New Orleans, almost completely destroyed by Hurricane Katrina, where houses are built on stilts; a First Nation settlement in the Louisiana wetland that is falling into the sea, its houses abandoned and refugees scattered; a submerged Roman party town in the caldera of a volcano; a rapidly sinking village on the coast of Wales near the site of a legendary lost city; an Atlantic archipelago threatened by saltwater intrusions and storm waves; drained former wetlands in England's East Anglia, formed from the mulch of a sunken forest, shrinking far below protective dykes as an ascendent North Sea pounds. These places are where the effects of global warming are most apparent – and most urgent – from rising sea levels to extreme rainfall and storm surges. They give us a stark glimpse of the future on a rapidly transforming planet. But they also tell a story about the unavoidability of change, the necessity of harmonising with nature's flux and the folly of trying to control it. In these threshold zones, water and land are embraced in eternal cycles of flood and reclamation, erosion and deposit, destruction and renewal. Forests become freshwater swamps. Swamps become saltmarshes. Saltmarshes become the seabed. The seabed becomes stone. Stone becomes the bedrock on which future forests grow.

Sunken lands show us that there have been many ends of the world, going back into the deep past. But the end is also a beginning, another turn of the wheel. Nothing really dies; it just changes. There have been many cultures who have endured global warming and survived the flood. If we peel back the layers of silt, sea and mythology, we can sometimes hear their ancestral voices and remember valuable lessons that have been forgotten.

2

THE FOREST BENEATH THE SEA

For as long as they could remember, they had lived by the river. One of several tribes in a verdant forest, they were tall and dark skinned with pot bellies, sated by the land that nourished them. They hunted red deer, wild boars and otters. They felled trees to build huts and boats. They collected berries, nuts and mushrooms, but only enough to sustain the web of life around them. They knew when it was the right time to pick, and the right time to let things grow; how much to cut and how much to kill. For all things were connected – the trees, the rocks, the plants – and to disrupt one was to disrupt all. Life was the water that flowed through everything, like the river through the forest, negotiating its eternal course.

One balmy evening they sat around the fire as musicians played tunes on bone flutes and beat a rhythm on hollow gourds. Stars sparkled in the black sky. The moon hung pale above the trees, strung out in heaven's high bower. All was bright and alive. But then they heard an unfamiliar sound: a thunderous cacophony of splintering wood amid an escalating roar. They looked around in panic. Some got to their feet and peered into the trees for signs of danger. Others drew their children close as the tumult swelled into a deafening crescendo.

In astonishment, they saw the river level plummet. For an eerie moment its exposed mudbanks gleamed in the glow of the campfire, writhing with silvery fish. That was all the tribe had time to comprehend before the river returned in a waist-high wall of foaming brown water. It smashed into their encampment in a torrent of branches, stones, moss, badgers, bears, cattle and deer; water-borne torpedo corpses that took out trees, huts and humans alike, swirling all living and non-living things together into a lethal soup and carrying it away through the sinking land.

It was difficult to picture a forest as I looked from the south coast of England over a sparkling sea towards a horizon busy with ferries and container ships. I was on the beach at Pett Level, a long arc of shingle, striated with wooden groynes, running from the foot of sandstone cliffs through a flatland of sheep towards the saltmarsh of Rye and the Camber sand dunes. Propping up the rear of the beach was the spine of a 3-kilometre-long earthwork defence where walkers stooped into the wind. To the east was the spit of Dungeness, its cuboid nuclear reactors silver against the blue sky, wind turbines spinning in the fields behind it. Under the silt before me lay forgotten habitats from a time when hunters stalked game through a forest of alder, willow, ash, birch and oak that ranged out towards France. The last of these trees were submerged three and a half thousand years ago but their stumps could still be seen embedded in clumps of peat on the shore at low tide.

My daughters and I were living on top of a sandstone promontory in neighbouring Hastings. It was a Tuesday morning in the first year of the coronavirus pandemic and the schools were closed. The girls were going stir crazy after weeks cooped up indoors, so I improvised a geography trip to the sunken forest. We piled into the car with our cocker spaniel Hendrix, and drove ten minutes east, where the road fell away to reveal a crescent of lowland, once a medieval tidal bay that turned to saltmarsh, later reclaimed for farming and holiday parks. A country lane took us through the village of Pett to the car park of the Smuggler's Inn. It had closed many years previously but there was still a sign outside offering a deal on burgers.

Despite the sunshine, the air was icy as we crunched on the shingle beneath sodden brown cliffs. Where the stones gave way to sand, we stepped onto a slippery expanse of peat. At first, it looked like rock, but when we kicked our heels into the surface, we revealed the malleable pitch-black sludge of the former forest floor. Occasionally we sank ankle deep into pockets of ooze. The intertidal actions of the sea had dredged tiny canals in the peat, through which water tumbled into foaming pools, littered with scallop and razor-clam shells. All around us were the remnants of trees; a woodland war zone as far as the eye could see: truncated oaks with splayed roots surrounded by toppled trunks, long as boat masts, draped in gutweed and bladderwrack. Some were skeletal ribs in the sand, while others – thick as a human leg – were strewn on the surface, as if a caber-tosser had gone berserk. Multiple thousands of pockmarks covered the mass of peat and wood: the work of the common piddock, clam-like creatures that burrow into self-created tombs, where they spend their whole lives, feeding on organic matter washed in by the currents, and illuminating the darkness with a blue-green light. When they die, they leave behind empty holes that become homes to other molluscs in search of shelter. The cycles of life and death, disaster and opportunity, revolved over many millennia in this landscape.

In his book *Submerged Forests* (1913) the palaeobotanist Clement Reid described how fishermen up and down the

coastline of Britain had for centuries told tales of tree stumps rooted in peaty earth, revealed between tides after storms. They were known as 'Noah's Woods' by locals, who assumed that these oaks and hazels were lost to the same flood that God unleashed on humankind in the Old Testament. Proof, surely, that what the Bible said was true. But the problem was, wrote Reid, that while some might explain them away with the notion of a biblical flood, 'it was difficult thus to account for trees rooted in their original soil, and yet now found well below the level of high tide'. To consider this was to throw out the chronology of the Bible and confront the dizzying implications of deep time, where the planet had experienced myriad geological shifts and sea-level changes going way back before human memory.

I showed my daughters that some of the wooden matter was soft enough to push a finger into. If you were to do so, I said, you might touch the same spot another child had touched long ago, before Jesus was born. These were part of a forest that grew here when Britain was not an island but joined to Europe by a landmass known as Doggerland, and when this part of the English Channel, just south of the Strait of Dover, was a river valley that had been gouged through chalk by a mega-flood 400,000 years previously.

The world looked very different at that time. Massive sheets of ice covered Canada, Scandinavia and northern Europe. With so much water locked in continental glaciers, the sea level was 125 metres lower, exposing coastal shelves, valleys and land bridges, traversed by giant deer, woolly rhinos and hippos. Across a tract of tundra from Spain to China, mammoths grazed on grasses and willow shrubs. Camels, giant beaver and bison wandered a steppe between Alaska and eastern Siberia. In the south-eastern woodlands of North America, lions and sabre-toothed cats prowled, while aurochs and red deer roamed the plains between Britain and northern Europe. Among these beasts were humans, hunting and migrating, adapting to the advance of ice and then to its retreat.

As the climate warmed, the glaciers melted and the seas rose. The withdrawal of ice removed an enormous weight off the land in Scotland and northern England. As the northern lands bounced back up, the south of England, which had been a bulge at the edge of an ice sheet, tilted downwards. This isostatic movement gradually sank Doggerland, year by year, while the ascendant seas flooded its hunting grounds and migratory routes, including the river valley in the Strait of Dover at its southernmost reaches. Then in 6200 BCE came the big one. A series of three submarine landslides off the coast of Norway triggered a tsunami that permanently flooded many coastal and lowland settlements around Britain and Europe, turning Doggerland into an offshore sand bank. Some historians believe that the Storegga Slides might have been the event that turned Britain into an island.

It is impossible to know whether some of the Stone Age people in the lowlands died violently, or if the inundation was gradual enough for them to flee. Either way, their homelands slipped beneath the water, never to be seen again – until 1931, when a North Sea fishing trawler dredged up a barbed harpoon made of red-deer antler. Since then, hundreds of objects have been raised from the seabed, including petrified hyena droppings, mammoth molars and a fragment of Neanderthal skull. Seismic surveys and digital modelling have revealed a complex topography of densely wooded valleys, marshlands and lagoons, inhabited by humans for thousands of years, then lost to memory – but not, perhaps, to the imagination. Long before its rediscovery, H. G. Wells conjured up a prescient vision of Doggerland in 'A Story of the Stone Age', published in the *Idler* magazine in 1897.

'This story is of a time beyond the memory of man, before the beginning of history,' Wells wrote, 'a time when one might have walked dryshod from France (as we call it now) to England, and when a broad and sluggish Thames flowed through its marshes to meet its father Rhine, flowing through a wide and level country that is under water in these latter days, and which we know by the name of the North Sea.'[1] Perhaps Wells was inspired by tales of submerged trees glimpsed off the English coast, or perhaps his imagination had intuitively accessed an ancestral cultural memory embedded in his DNA. This might sound outlandish, but animal studies have shown that experiences of fear and trauma can be passed down through generations in a form of genetic memory, a process known as 'transgenerational epigenetic inheritance'.[2] Each one of us embodies the accumulated behavioural modifications of our species, and those other hominids which came before us and bred with us when we co-existed on the planet. The human genus has evolved through radically changing climates and ecosystems, adapting new behaviours after confronting new existential challenges, of which the palaeo-tsunami that transformed Britain was a very recent one in its 2.8-million-year history.

In the English Channel, 160 kilometres west of Pett Level, archaeologists have discovered artefacts contemporaneous with the Storegga Slides. In 1999, divers surveying the seabed near the Isle of Wight saw a lobster digging up manmade tools. Excavations revealed evidence of a settlement dating back 8,000 years, including flints, string, hearths and burnt wood. Divers from the Maritime Archaeology Trust have since found split timbers and the wooden foundations of a circular building, indicating the use of crafting technologies far beyond the capabilities expected of that time. They have declared it the world's oldest boat-building site and the 'most cohesive, wooden Stone Age structure ever found in the UK'.[3] Further studies around the Isle of Wight site have found the DNA of einkorn wheat. This variety was dominant in the eastern Mediterranean 10,000 years ago, which means that this was an imported crop, brought to the south of England on boats over huge distances.[4] Therefore, the people who experienced the flood in 6200 BCE were more sophisticated and worldly than we might imagine.

The tribes living in the forest valley now under the English Channel would have made it their home, coppicing and burning to encourage the growth of nutritious hazel trees and edible plants, filling it with ancestors and spirits, storying the landscape with myths, constructing tools, dwellings, structures for worship and boats. But as global warming continued to reshape the world, the encroaching sea forced them to migrate to higher ground. Layers of silt encased their abandoned settlements in a preservative, oxygen-free sludge for thousands of years, allowing twenty-first-century researchers to touch the remnants of campfires as if they were extinguished only yesterday.

As I gazed out over the sea from Pett Level, I thought about the artefacts that might be concealed in mummified chambers on the seabed and imagined what their owners might think if they were reanimated, suddenly, only to see mackerel, jellyfish and P&O ferries soaring above their village. Could they conceive of their world's annihilation? I imagined myself hurled forwards thousands of years to find everything I knew under water. My house. My road. My town. Barnacled cars, crawling with starfish. Basking sharks above the rooftops. Currents carrying away my bloated possessions. Perhaps if I were to swim to the surface I might see a shingle shore, where a super-evolved mutant author scribbled notes for his book, *Sunken Lands*, wondering what happened to the people who once lived beneath the sea. Like H. G. Wells and me, millennia before him, the mutant author couldn't possibly know what those humans felt, or how their lost civilisation functioned. The only way to access that past was through his imagination, and an attempt at empathy with those ancestors who plundered the world to the brink of extinction.

The inundation of their land must have been perplexing for the people of the forest in Pett Level, as great pools of ground-water expanded, trees died and their land turned to marsh. But it would take thousands more years for their world to vanish. Life remained in the dwindling forest until the last trees were submerged in 1500 BCE, when sea levels stabilised. In 2014, a team led by geoarchaeologist Dr Scott Timpany collected samples from the beach and found evidence of many different incarnations of the forest from the Mesolithic through to the Bronze Age. At one time there was a ground covering of water avens, pond weeds, greater spearwort, common bulrush, marshwort and sedges. The woods were a saltmarsh for a period before a return of ash, willow, birch and hazel, with elm, pine and whitebeam on the clifftops where it was drier. They found deadwood fungi in rotten branches, a hazelnut gnawed by a mouse and charcoal from manmade fires. Based on these discoveries, the artist Alice Watterson painted a reconstruction: a majestic stag standing among lush green trees and shrubs beneath the sandstone cliff, the only feature remaining today, though it is eroding at the rate of 25 metres a year and won't last forever. The village of Fairlight is on top of the cliff. All that stops it from falling into the sea is a barrier of granite boulders.

Pett Level, too, is under threat. Across the beach, wooden groynes are piled with stones swept east by the force of longshore drift, dragging at the southern coastline of England. Storm tides are testing the sea defence, formed from a manmade embankment at the rear of the beach that once bore the railway line that carried displaced shingle back to Pett Level from its accumulation at Rye. Behind it is a terrace of coastguard cottages, whose residents' rear-window view is no longer of the sea but of a steep slope of grass. However, the newer houses have been built to rise above the earthwork in an array of experimental modern styles, the most famous being the Bauhaus-inspired white cube owned by the late Tom Watkins, former manager of the Pet Shop Boys. But no matter how grandly designed, all that protects them from inundation is a mound of earth; once that is breached, they will be left at the mercy of an unforgiving sea. It might not be long before the White Cube becomes another intertidal spectre, pitted with piddocks, coated in green slime and dripping with hubris.

After all, this has happened here before. At the far end of the sea-defence line, the sand dunes of Camber were a crest of white curving towards Dungeness. Somewhere near this spot the seaport of Old Winchelsea stood at the edge of a shingle spit. Its prominence in the English Channel had once made it an international trading hub, with French wine delivered on ships to its bustling quay. In the eleventh century it became a Cinque Port, part of a confederation whose members were given economic perks in return for defending the south coast on behalf of the

Crown. But by the thirteenth century, there was trouble afoot. A period of global warming that began in 900 CE, raising sea levels, had begun to reverse, in what is known as the 'Little Ice Age'. Plunging temperatures brought savage storms to Europe. From 1233 onwards, these tempests began to inundate the winding streets of Winchelsea at high tide. To make matters worse, long-shore drift was undermining the shingle promontory on which the port was built. Then, on 1 October 1250, an ominous blood-red moon heralded a storm described by the chronicler Raphael Holinshed as 'so huge and mightie, both by land and sea, that the like had not been lightlie knowne, and seldome, or rather never heard of by men then alive'. He described how the sea surged through the town, destroying 'bridges, mills, breaks, and banks, there were 300 houses and some churches drowned'. Two years later, another storm struck, its devastation recorded this time by Matthew Paris, who described how the waters 'drowned and washed away a great many of the inhabitants'.

For thirty more years, the townsfolk pleaded to the Crown for help as the shingle bank disintegrated beneath them. King Edward I was on the throne, an unusually tall man with a penchant for planning new towns. He understood that it would not be pos-sible to save Winchelsea from the sea. Instead he would build a new version on the hill behind the saltmarsh, where an estuary allowed ships to offload their goods at high tide. His architects started work on the designs in the 1280s, while the citizens of the dying town awaited the inevitable. In 1287 a final storm struck the fatal blow, killing hundreds of people and animals, destroying its buildings, and leaving the remains permanently inundated. The coastline was radically reshaped overnight, the Rother estu-ary now lapping at the edges of Rye, a formerly landlocked town, and most of the Norman castle in Hastings lying in the surf below the cliff. Over the years, Winchelsea's remains were carried off by currents, or bogged down in silt, there to join the Stone Age forest beneath the channel.

The medieval climate-change refugees of Winchelsea would have had no conception of the Mesolithic flood disasters preceding their experience – any more than those Stone Age people could conceive of the topography that existed millions of years before their valley was even created. Changes to sea levels are but minor fluctuations within the colossal ebb and flow of geology, where lava is spewed from the mantle to create new crust, and tectonic plates smash into each other, producing mountains that slowly erode into sediment that is pressed into new forms of rock. Organic life decays into layers of matter, slowly turned to stone that will one day become the solid ground on which new life grows, before it too is turned to stone. And so the cycle turns. Before the Mesolithic forest there were other forests, and before the beach, other sands when dinosaurs dominated the earth. In the Cretaceous, the foreshore below the cliffs of Fairlight was the bank of a lagoon, fed by many rivers, into which sluiced fragments of plants and dead animals, covered in silt and compressed into sandstone. Iguanodon prints have been found among the bones and teeth of freshwater sharks, crocodiles and turtles. After hundreds of millions of years beneath the ground they were forced back into the sunlight by tectonic uplifts, eroded by water and wind, for our civilisation to discover. They inspired revolutionary ideas in nineteenth- and twentieth-century thinkers who realised that the world was far older than people imagined and that strange monsters lived and died long before God was presumed to have gifted the planet to humankind as their sole domain.

One of these thinkers was the French Jesuit priest and palaeontologist Pierre Teilhard de Chardin, a student at the seminary in Hastings in 1908, and a keen fossil hunter who came regularly to this foreshore. As he perused the remains in layers of exposed

Ashdown Sandstone he began to consolidate his scientific and theological knowledge into a theory that these ancient creatures, and their extinction, were part of a long evolutionary process in which nature became increasingly complex. Humans were the greatest leap forward yet. The development of consciousness had created a new layer of philosophy, science, culture and technology over the biosphere, which in 1922 he termed the 'noosphere'. Humans had not simply evolved. We had become the agents of evolution, with the potential to transform ourselves and the world around us. Years later, he expounded his theories in his book *The Phenomenon of Man* (1938), in which he wrote:

> This sudden deluge of cerebralisation, this biological invasion of a new animal type which gradually eliminates or subjects all forms of life that are not human, this irresistible tide of fields and factories, this immense and growing edifice of matter and ideas – all these signs we look at, for days on end – to proclaim that there has been a change in the earth and a change of planetary magnitude.

Teilhard's metaphor for human expansion was one of a rising tide; a deluge. But rather than destructive, he believed this flood to be spiritually necessary. Our proliferating technology was a medium for the advance of consciousness, which would keep accelerating until we transcended the material realm and converged with God. He called this the Omega Point. The great singularity. Some believe this is fast approaching, thanks to artificial intelligence, in which a cyborg fusion of human and computer minds might soon make scientific, economic and social changes at a pace that rapidly transforms our reality. He was the godfather of the kind of accelerationist, utopian thinking that demands free-market capitalism keeps pushing technology further and further, faster and faster, no matter the environmental or social cost, so that humanity might be liberated from the earthly chains of time and space.

This is typical of a mindset that posits the planet as divinely gifted material for the spiritual pursuit of humanity; that we are entitled to manipulate the world like gods; and that we progress inevitably towards enlightenment. But the dead trees on the beach of Pett Level suggested that history was cyclical, rather than linear. Creation and destruction came in recurring cycles. Forests grew and forests died. Species flourished, then vanished. Thriving cultures became forgotten. Who was to say that the folk who once lived in the forest were less complex, or less close to the divine, than we are in the twenty-first century, addicted to cars, smartphones and fast food, embattled by carcinogens and enslaved to fossil fuels? Without the biosphere, with its delicately interconnected systems, there could be no 'noosphere'. Progress was not inevitable. After all, this ancient woodland was a place that people believed would last forever and now it was gone.

Another famous person inspired by this layered landscape was David Bowie. In 1980, he and director David Mallet chose Pett Level as the location of the video for his single 'Ashes to Ashes'. The lyrics referenced Major Tom from his 1969 hit 'Space Oddity' and suggested that Bowie was putting to rest his former personas, rebirthing himself as someone new. The video would be a lavish manifestation of this concept, in which he would lead a bunch of New Romantics across the beach in a funereal parade. Among frazzled shots of the sunken forest, craggy cliffs and golden shingle, there are scenes of Bowie semi-submerged in the shallows as if awaiting baptism, and others in which he is strung up in a womb-like chamber. His work was all about change, shedding skins and adapting to shifts in culture, so Pett Level was the perfect place for his Doctor Who-style regeneration, defined by cycles of destruction and renewal, where forests turned into marshes, marshes turned into fields, fields turned back into marshes, and the marshes turned to sea. Each disappearance led to a new emergence.

Thirty-five years after Bowie's metamorphosis on Pett Level, I hit an all-time low when my marriage ended. A shadow had been looming across our relationship for months, the dark spectre of a threat I had failed to investigate. One night, as we thrashed out our problems across the kitchen table, she looked into my eyes and announced that it was over. The next morning, I left as soon as I could with no destination in mind. I walked to the edge of town, where the road gave way to sandstone cliffs. Then I clambered up steep steps to the top of the cliffs where I paused for breath on a crest of green, ringed by golden gorse, high above the English Channel, with Hastings tucked in a valley behind me and hills unfolding towards Dungeness. I made my way along a twisting coastal path that eventually plunged into the flatland of Pett Level. I sat in the Smuggler's Inn and downed three pints of beer before launching myself onto the beach, determined to walk to Winchelsea, despite the deteriorating weather. I felt raw and exposed on the ridge between the shingle and marshland, wind roaring in my ears. It was as though my skin was being torn from my face. To hell with it. Let it all come off, along with those things that were no longer to be – the Christmases in the home we'd renovated, the garden we'd planned, the Sunday dinners and pub roasts, visits from the kids as young adults, growing old together, laughing at shared memories. These images peeled from my mind, whipped behind me like the torn pages of a picture book in the wind.

The path was so long and straight, it felt as if I was walking against the direction of a travelator. Dungeness power station seemed always the same distance away, each groyne a repetition of the one before, every sheep field a replica. After a while, my exhausted mind wandered out to sea, where the surf crashed over the dinosaur-bone lagoon, the Stone Age forest, the town of Old Winchelsea and the wreckage of ships. I felt tiny and unimportant next to all that history. My experience only one of many, reaching back into the monstrous chasm of deep time, a black hole of forgotten memories. This dizzying blast of perspective was of some consolation. The topography of Pett Level was one of turning tides, shifting banks of shingle, emerging trees and crumbling peat beds. Nothing stayed the same in this threshold zone, not even for a moment. Sand, water, rock, cloud and wind were in a perpetual entropic dance of erosion. Like so many others before and after me, I would have to accept my loss, suffer the pain, adapt and start again: my own microscopic contribution to the eternal cycle of events. Life was flux. Change was inevitable.

And now here I was on Pett Level, three years later. My daughters were behind me on the shingle, taking a break from my incessant gesturing at dead trees, while Hendrix furtively pawed for something hidden beneath the stones. Since that anguished walk I had joined a rock band, written music, published books, moved house, started a new relationship and made friends I'd never have otherwise met. A different reality had emerged from the withered husk of the old one. I was emotionally healed but scarred nonetheless. Even when a trauma passes, it leaves welts in your psyche, like the footprints of extinct monsters in fossilised silt. By the time we reach middle age we are lumpy with accumulated experiences. The longshore drift of life. But this is no bad thing. It can make us kinder and wiser, more appreciative of our miraculous existence in a fragile, transient world.

I looked out to where the surf lapped the sand on the far side of the sprawling mass of peat, logs and seaweed. A bald man in swimming trunks paced to and fro in a curious, stooped manner, scanning the length of the beach. After a while, he beckoned to a passing child and spoke to him. I couldn't make out the words, but his tone was angry. The little boy shook his head and made his way back to his mother, who stood with a small dog on a sand bank, staring at her phone. Seemingly enraged, the swimmer pivoted and continued his march up and down the edge of shattered forest as if trapped between land and sea in some infernal loop. Curious, I made my way down to the ridges of peat. I could hear the boy say something to his mother about a lost bag. That's when it clicked. The man was looking for his dry clothes. After being pulled this way and that by the current, he had emerged to encounter an unrecognisable morass of wood, peat and sand. Here the land and sea are in constant motion and never the same, even within the space of an hour, with very few landmarks with which a person can orientate themself. Eventually, he bent down and grabbed an object – a brown satchel, almost the same colour as the sand – and held it aloft triumphantly, like the severed head of a guillotined king, for the benefit of the disinterested few on the beach.

I turned to see my daughters and dog beneath the ridge I had stumbled along three years before, embroiled in my own private crisis. Did anyone observe me on the path from a distance, as I had the sea swimmer? Did they notice my agonised facial expression? Did they wonder if I was in trouble? That day felt so long ago, and as if it were yesterday. Those sedimentary cliffs with their skirts of scree, stark against the sky, appeared no different than they had in 2016 or when they were the backdrop of Bowie's 'Ashes to Ashes' video, despite the constantly crumbling sandstone and renegotiations of sand and shingle. At this thought, the track's haunting synthesiser refrain, warbling like a warped cassette tape, ear-wormed into my brain, dissolving the gossamer threads of time. I could see my heartbroken self on the pathway, while David Bowie's entourage marched in a funeral procession for his previous personas, preparing for his imminent rebirth. On the rocks, Teilhard de Chardin searched for God in the fossilised silt of a Cretaceous lagoon, picking through the bones of an idea that would make him famous. And on the cliff, a Mesolithic hunter watched over the forest canopy for rustling prey, listening to the cries of his comrades below.

In my mind's eye, we motley crew shared the same storyline. Because although the floods that drowned the forest seemed like distant history, in geological terms they happened recently. The Mesolithic hunters, Teilhard de Chardin, David Bowie and I were fellow denizens of the Holocene, a period which accounts for less than 0.1 per cent of the history of life on Earth. We coexisted in what musician Brian Eno calls the Long Now, which we will also share with those humans to be born in the next thousand years. Eno wrote: 'The Long Now is the recognition that the precise moment you're in grows out of the past and is a seed for the future. The longer your sense of Now, the more past and future it includes.'[5]

Before my children were born, I knew of the many existential threats to humanity, but they seemed abstract predictions for a distant future. I didn't consider that my offspring might experience the collapse of civilisation before they reached my age.* But climate change has escalated quickly in the last decade, with smashed records for global temperatures, carbon-dioxide levels and rates of ice loss. Sea levels are rising faster than they have done for 3,000 years. The UN secretary general has warned of 'a mass exodus of entire populations on a biblical scale' with entire nations lost to the waves.[6] The challenges of the Stone Age world are now our challenges too, as they will be for descendants we

---

* In 1972, a study by the Massachusetts Institute of Technology predicted an industrial worldwide collapse by the year 2040.

will never know, but for whom we are responsible through our actions today.

This all became starkly apparent on the beach at Pett Level, the epitome of the Long Now, where I pressed my toes into 4,000-year-old peat within sight of a nuclear power station that will remain toxic for millennia beneath the sea that will one day overwhelm it. Here, the epochs of the past and future are interwoven, layered with multiple thousands of experiences that occurred in the land's many incarnations, from forest to marsh to sea to reclaimed land. A red deer, staggering at the thud of an arrow. A poisoned tree sinking into the salty mire. Screams from the victims of a medieval storm. A flicker of blue light from a piddock in its burrow. A parade of New Romantics, their heads bowed in reverence. A sea swimmer in a desperate search for his clothes. These were the intricate emotional resonances that trembled beneath the visible and transcended the temporal. They told a story about this place beyond that which had been written, recorded or even remembered. One that could be divined by imaginative voyages, not only into the past, but forward into storylines that might unfold after this beach is once again beneath the sea, our designer homes, smartphones and bones entombed in silt.

Eventually, after thousands of years of submergence, these compressed remains might become thrust back into the light in this same spot, water-weathered, weed-slimed and crustacean-encrusted; weird objects from a forgotten time to confound minds and inspire new ideas about who we are, where we came from, and where we might be destined to go.

3

LOST KINGDOMS

In the West of Wales there was once a kingdom named Cantre'r Gwaelod, ruled by King Gwyddno Garanhir. Sixteen settlements were connected by roads in a fertile lowland of woodlands, rivers and grassy plains. It was said that an acre of field there could produce as many crops as four acres anywhere else in Wales.

The wealthy king enjoyed the finer things in life, and lived in a fortress where he entertained his nobles with raucous parties. In the courtyard was a freshwater well, attended by Mererid, the cup-bearer who brought jugs of water to the king and poured him his wine. She was the people's conduit to the deities and would plead on their behalf should the well dry up or overflow.

They had good reason to worry. Every year, the rising sea between Ireland and Wales took more of the lowlands, mile after mile, turning forests to swamp. Violent storms brought closer the turbulent brine until it began to inundate the Cantre'r Gwaelod, killing crops and destroying houses. To protect his citizens, King Gwyddno ordered the construction of a 3-kilometre-long dyke. Its sluice gates would open at low tide to drain the farms and woodlands, then close before high tide. He entrusted responsibility for the gates to Seithenyn, the most intelligent of his noblemen.

Each day, an increasingly melancholic Seithenyn sat at his post, high on the wall, and saw the world change: forests sinking into the ooze; trees staggering in the saltmarsh; glistening pools where leafy canopies used to be; viscid green tide lines on the sea wall inching higher and higher. Yet the king and his noblemen ignored his warnings and continued to live as they always had done, believing that the wall would protect them.

One fateful day, King Gwyddno held a grand feast. Seithenyn was ordered to attend. But all through the evening he grew nervous about the dark clouds overhead and the high winds that blew. After consoling his troubled mind with wine, Seithenyn slunk to his post, where he fell into a deep drunken sleep, only to be awoken hours later by the cries of Mererid from down below.

'Seithenyn! Look out at the waves crashing over Gwyddno's realm!'

He looked down to see the dyke's gates wide open and the sea rolling in at high speed. Water crashed through the courtyard in a foment of flailing arms and legs, tables, chairs and roasted meats. Mererid was clutching onto the crank of the holy well with both her hands, the sea up to her chest, pleading with the gods to stop the flood, until she could hold on no longer and vanished from sight.

The terrified king and his nobles emerged from their balconies and frantically scaled the walls. Seithenyn ushered them onto a causeway, named Sarn Cynfelyn, which led to higher ground. As the bedraggled survivors fled for their lives, Siethenyn looked behind him to see a towering wave fill the sky, before it came crashing down, submerging Cantre'r Gwaelod forever.[1]

Rain hammered down so hard on the M6 motorway it felt as if my car roof might buckle and crush me in my seat. I was leaving the gravitational field of Birmingham, its sprawl of factories, flyovers and industrial parks fading behind me, and the north-west somewhere beyond a veil of cloud. The road was a tumult of roaring engines and hissing tyres. Raindrops sparkled in headlamp beams. Muddy spray, flung from truck wheels, fizzed in chaotic Brownian motion. My wipers thrashed dirty rainbows back and forth across the windscreen as water snaked along the rubber seams, as if probing weak spots for entry. The deluge was horrible to drive through but felt highly appropriate to my mission: to search for submerged Welsh kingdoms, sunk by climatic cataclysms and the follies of humankind.

My father's side of the family is Welsh and I visited them regularly as a child. We'd drive to my nana's house in Wrexham for Sunday lunch, then make a short hop to the village of Bangor-on-Dee where two of my great-aunts had a farmhouse near the racecourse, and a third great-aunt – Bette – lived in a cottage next to the post office. They had long since passed away, and I had not been back to Bangor-on-Dee for many years, but it was to be my first destination as I crossed the border from England into Wales, for the village was deeply afflicted by floods, thanks to its troublesome River Dee.

The rain had stopped by the time I pulled up beside the cottage where Bette taught me how to play backgammon, smoking cigarettes while her ageing whippet snoozed beneath a blanket. Bangor-on-Dee was nestled on a bend in the river, spanned by a five-arched seventeenth-century sandstone bridge. Its church, St Dunawd's, was behind an embankment fortified by gabion bulwarks, netted and tied with metal rings to protect it from the rising river. Beyond the opposite bank, a raised road was holed with tunnels to let water surge through to the floodplain. The church was open, but its lights were off, and the interior was gloomy. Historical information boards lined the aisle beside an animatronic monk at a writing desk. I flicked on my smartphone torch and its beam alighted on a sepia photograph of two elderly ladies in raincoats and headscarves, knee-deep in floodwater on Station Road. It was as if I'd caught them by surprise in their moment in history. My dad recalls a photo of his grandmother hoisted on a chair above muddy waters in the doorway of his aunts' house and I wondered if these women were her contemporaries. Adjacent to their photo was a black-and-white aerial shot of the village submerged, its ramshackle stone bridge looking like a diplodocus skeleton sinking into an eternity of silt.

The threat had grown more severe over the decades. A drainage system, designed to force surface water back into the Dee, was struggling with record river levels and downpours. A few months before my visit, homes had been evacuated during a storm when the Dee had breached the defences. Aerial footage showed HGV trucks like barges cruising down canalised roads and the sailboat masts of telegraph poles jutting from swirling currents. That same year, 80 kilometres west of Bangor-on-Dee, the River Conwy had overflowed, turning the gardens of Gwydir Castle into a lake, local roads into rivers and farm fields into open water, rippled with waves. These scenes were such a regular occurrence that the Welsh government had declared a national climate emergency.

As I left the churchyard and walked along the rise of the fortified riverbank earthwork, I looked across the rooftops and imagined Auntie Bette's cottage one day vanishing beneath the surface of the expanding Dee until all that remained was the church tower, looming above inky waters, its ghost bells ringing for centuries after its congregation had been forgotten. Should this be its fate, Bangor-on-Dee would join the many submerged towns and villages in Welsh folklore, like the one at the bottom of Llyn-y-Maes in Cardiganshire, sunk as punishment for its debauchery; or the one in Kenfig Pool, Glamorgan, drowned after its mayor fatally wounded a visiting prince, who cursed them with his final breath; or the one that supposedly lies under Llyn Syfaddan, also known as Llangorse Lake, glimpses of which have been reported for almost a thousand years.

In 1188 CE, the archdeacon and historian Gerald of Wales spoke to the locals of Llyn Syfaddan, who told him they experienced visions of the lake 'adorned with buildings, pastures, gardens and orchards'. The waters would turn green overnight. Mysterious red channels appeared, as if blood flowed through invisible veins. When it iced over in winter, they heard groaning from its depths. In 1586, William Camden wrote of an 'ancient tradition that where the lake now is, there was formerly a City, which being swallowed up by an earthquake, resigned its place to the waters'.[2] These two accounts were almost four centuries apart, showing how enduring this legend was for local people. It may have been a trace memory of ancient communities who lived on a crannog, an artificial island on the lake, and might have been affected by climate change.

At the end of the Bronze Age, there was a 300-year-long dry period, followed by a return of the rains. Evidence shows that a geologically similar neighbouring lake dried out then flooded again quickly as the Iron Age commenced. Swords, sickles, cauldrons and axes have been found in its depths, suggesting ritualistic appeasement of the ominously rising water.[3] Perhaps this phenomenon also happened to the people who once lived in Llyn Syfaddan, and the ghost of their experience lingered on in stories for centuries afterwards, mutating into new narrative forms as it drifted through eras with different cultural concerns.

By the nineteenth century, there were numerous folkloric versions of how the town sank. One published in *The Mythology and Rites of the Ancient Druids* by Edward Davies (1809) told of a decadent populace who loved to party. When rumours of their behaviour reached a local king, he sent an emissary to find out what was going on. This royal agent was ignored by the villagers, who were embroiled in Hogarthian binge-drinking and had no interest in his questions. Looking for someone sensible to accommodate him, the emissary entered a cottage. Nobody was home, except for a baby in a cradle. He tarried a while, staring bleakly at the poor child, then departed, not realising he had left behind one of his expensive gloves. On his way back to the king, the emissary heard crashes of thunder, with screams and shouts. At that moment he realised that his glove was missing and returned, only to find the village replaced by a lake. There was no sign of life but for an object adrift on the surface. It was the cradle, containing the abandoned baby: the sole survivor of the disaster.*

Most tales of sunken Welsh villages blame the sinful behaviour of their inhabitants for their destruction. As far as I was aware, Bangor-on-Dee was not a hotbed of deviancy but, all the same, one day a great flood might come, the pumping system might fail and the village of Bangor-on-Dee might sink, just like the Welsh towns of legend. Its memory might survive only through my telling of its story, and whoever might retell it after me, embellished and refined into a morality fable for future times.

Wales is a nation soaked in flood mythology. This might be because its western regions were shaped by an expanding Irish Sea that inundated inhabited lands and hunting grounds over many thousands of years. At the peak of the last Ice Age, Wales was joined to Ireland beneath a kilometre-thick glacier. As it retreated, a vast proglacial lake formed. According to author Philip Runggaldier, this lake was breached 14,750 years

---

* This version is also the one later told by Marie Trevelyan in her book *Folk-lore and Folk-stories of Wales*, published in 1909.

ago, unleashing a mega-flood that swamped West Wales and the south-west of England, killing two-thirds of the Palaeolithic population who lived there. This could be the source of the Welsh legend of Llyn Llion or 'the lake of floods'. It appears in the Welsh Triads, medieval texts of folklore, mythology and history collated (and embellished) in the early nineteenth century by the poet Iolo Morganwg. He wrote of 'The three burstings of the Lake of Llion: the first, when the world and all living beings were drowned, except Dwyvan and Dwyvach, their children, and grand-children, from whom the world was again peopled – and it was from that bursting that seas were formed; the second was, when the sea went amidst the lands, without either wind or tide; the third was, when the earth burst asunder by means of the powerful agitation, so that the water spouted forth even to the vault of the sky, and all of the nation of the Cymry were drowned, except seventy persons, and the Isle of Britain was parted from Ireland.'[4]

The settlements lost to soaring sea levels throughout the early Holocene might have given rise to the nation's most famous legend of a sunken land, which opened this chapter: the kingdom known in Welsh as Cantre'r Gwaelod, and in English as the Lowland Hundred, which now lies beneath the Irish Sea. The story is about a network of sixteen towns on a fertile plain, ruled by King Gwyddno Garanhir from his fortress, Caer Wyddno, where a holy well was protected by the cup-bearer Mererid. To protect it from the sea, the king built a dyke and gave responsibility for its flood gates to a noble named Seithenyn. But one fateful evening it was sunk beneath the waves after a great feast.

The legend was first recorded in the thirteenth century, where it took the form of a poem entitled 'Boddi Maes Gwyddno' ('The Drowning of the Land of Gwyddno') in the *Black Book of Carmarthen*, the earliest surviving manuscript written in Welsh. It begins:

> Seithennin, stand forth
> And behold the seething ocean:
> It has covered Gwyddno's lands.[5]

The book brought together legends that had been circulating in oral lore for many centuries before they were committed to text. This story is believed to be an account of events in Welsh legend that purportedly happened in the fifth or sixth century. But its roots probably reach much further back to when sea levels were 30 metres lower and a forest filled what is now Cardigan Bay on the west coast of Wales, a 65-kilometre-long crescent that – on a map – looks as though a chunk has been bitten out of an apple. A lowland stretched across the bay from the town of Cardigan, in the south, to Bardsey Island, which hangs off the crescent's horn to the north. During the global warming of the early Holocene, thermal expansion and glacial flooding swelled the Irish Sea, incrementally submerging the flatland. Geoscientist Patrick Nunn has described how peat sediments on the seabed have enabled scientists to track this process of post-glacial sea-level rise.[6] If their analysis is correct then the last time dry land connected Cardigan to Bardsey was at least 9,000 years ago. Therefore, it might be from this period that the story originated, when recurring floods forced the forest's inhabitants to abandon their ancestral homes and migrate to higher ground, until all that remained of their abandoned world were stumps of birch, pine and oak, visible at low tide on the beaches of Ynyslas and Borth, along with local lore about a lost civilisation.

These legends gained credence from the coastline's post-glacial topography, which featured uncannily linear accumulations of stone left behind by the retreating ice, known as terminal moraine. These rock formations were hard to explain in an age before we understood the deep-time geological actions of ice, and were therefore assumed to be effects of human agency. In Cardigan Bay terminal moraines in the form of three reefs extended from the shore at right angles. They became known as Sarn Badrig, Sarn-y-Bwch and Sarn Cynfelyn and were interpreted as causeways to the Lowland Hundred. In the seventeenth century, Robert Powell Vaughan believed that Sarn Badrig was 'a stone wall made as a fence against the sea'. Antiquarian William Owen Pughe described how, as a boy, he sailed over the ruins of habitations in Cardigan Bay, 5 or 6 kilometres off the coast. 'Many of the stones seemed to be large slabs,' he wrote, 'and lying to confusion on the heap.' Another cluster of erratic boulders, 11 kilometres out, seen only at very low tide, was believed to be the remains of the king's fortress, Caer Wyddno. It was said that its bells could sometimes be heard from the beach of nearby Aberdovey on the edge of the Dyfi estuary, where water from Snowdonia flowed into Cardigan Bay.

In the eighteenth century, a popular song based on the legend, 'The Bells of Aberdovey', was sung in music halls and, in 1936, a new chime of bells was installed in St Peter's Church to let its melody soar across the village and over the water. Almost a century on, the Lowland Hundred continues to influence artists and musicians. Beneath a wooden jetty in Aberdovey's harbour, a bell installed by sculptor Marcus Vergette hangs over the water, with a clapper made from carved oak to catch the rising tide, sounding the toll for the drowned. In 2008, Paul Newland and Tim Noble, musicians from Aberystwyth, a town in Cardigan Bay, named their band the Lowland Hundred and performed shimmering, glacially slow songs about loss and memory. 'The present wears half-remembered dreams and memories,' sang Newland, 'Lichen grows on the rocks. Broken oak.' More recently, musician Jim Finnis uploaded a live stream of ambient music inspired by the Lowland Hundred, which fluctuated with the motions of the sea, rising to a climax of horns and bells at high tide, then waning into drones at low tide. It is amazing how prehistoric events at the end of the Ice Age, witnessed by a forgotten people, can persist through oral lore, medieval myth and Victorian music hall to alternative rock and electronic ambient live streams, their ancient waveforms oscillating through optic cables and 5G networks, emerging from computers and phones around the globe. It is as if we are collaborating with our ancestors in a song in the Long Now, as relevant as when the Irish Sea surged over Mesolithic settlements. Through myth, the past speaks in the present, sounding a warning that must be heeded.

On a chilly afternoon, I arrived at Borth to see King Gwyddno's land for myself. I parked at the sea wall and hopped down a shingle bank to the beach. In 2019, Storm Hannah battered Cardigan Bay with 130 kilometres per hour winds, scouring the sands, revealing more of the prehistoric forest than had been seen before. Thousands of stumps, some jagged sharp like fins, some bulbous like skulls, ranged across the sands and poked from foaming shallows. Whereas on Pett Level the remnants were scattered branches and logs, here the stumps were evenly spaced, a few yards apart, all roughly the same height. It was as if a nuclear bomb had just blasted through a living forest. Much of what I could see had been encased in acidic peat until a year earlier, so the wood was miraculously preserved. I meandered between stumps, marvelling at the nut-brown swirl of the grain and scabs of bark. The welts in the peat beds looked like giant footprints and, indeed, the impressions of bears, sheep and humans had been found here, including one of a four-year-old child. Archaeologists had also discovered an antler tool, flints, burnt-out hearths and the skeleton of an auroch.

As I stepped from branch to branch, avoiding the boot-sucking sand, I wouldn't have been surprised to see horns jutting from the soggy morass, or a clawed human hand clutching a flint. For it felt as though more revelations were imminent, that the forest wasn't sinking but growing, coming back to life. Some of the oak stumps were coated in hair-like seaweed with big roots splayed out like arms on each side as if they were zombies of the Lowland Hundred, hoisting themselves out of the silt to warn their descendants about history repeating. For the same environmental forces that destroyed the land of Gwyddno are about to destroy the village of Borth. It is estimated that sea levels on the west coast of Wales have been rising by 5 millimetres every year, but this process is now accelerating exponentially.[7] Predictive maps produced by the international charity Climate Central show that many of the saltmarshes and mudflats around Cardigan Bay could be under the sea by 2030, while low-lying villages such as Borth could sink within a decade. When storm Barra struck at the end of 2021, waves topped the sea wall and flooded the streets. One resident, Kim Williams, told ITV Wales, 'We need help with storm boards or storm gates but this is just a plaster on a wound. It makes people feel better but the problem isn't going away.' She, like many others, is beginning to understand that we cannot stop the sea rising. Neither can we ignore it. 'We have to be realistic,' she said. 'Things are going to change.'

Whatever the origins of the Lowland Hundred legend, they were long before the tale was written in the *Black Book of Carmarthen*, and long before its fifth-century setting, back when the dead forest I was walking on was alive, thick with trees, noisy with animals. In his paper to the Cambrian Archaeological Association in 1849, Griffith Edwards proposed that whatever lay beneath Cardigan Bay could not have been submerged in the fifth or sixth centuries but was likely to have been pre-Roman. Later that century, Professor William Dawkins proposed that an ancient people travelled and traded between the Iberian

Peninsula and Wales. 'A forest, containing the remains of their domestic oxen that had run wild, and of the indigenous wild animals such as the bear and the red deer,' he wrote, 'occupied the shallows of Cardigan Bay, known in legend as "the lost lands of Wales".'* The people who migrated up and down the Celtish coastline between Portugal and Wales may have experienced the same flooding of their lands or shared stories with each other.

This might explain the similarities between the Lowland Hundred story and Brittany's legend of Ys, a trading port in the Bay of Douarnenez that was notorious for its decadence. Vulnerable to high tides, it was protected by a dyke with a sluice gate, the key to which was in the king's possession. It was foretold by Saint Gwenole of the Landévennec monastery that the sins of Ys would lead to their destruction. The king was a pious Christian who took counsel from Gwenole. He wholeheartedly agreed that his town must change its ways. However, his daughter, Princess Dahut, was one of the main culprits – a woman who enjoyed sex, heaven forbid. She ignored her dad's demands

---

* From his 1899 address to the university in Bangor.

of chastity and snuck out one stormy night to meet a lover, an irresistible red knight who was arriving from out of town. She stole the king's key and unlocked the gates, not realising that the tide was high, nor that her boyfriend was, in fact, the Devil, and she had been tricked. The water poured in and Ys began to flood. Swooping through the torrent on his horse, the king rescued his daughter and headed for higher ground. But the voice of Saint Gwenole in the wind told him that he must cast his demonic daughter into the sea if he was to escape with his life. After the ghastly deed was done, Ys disappeared beneath the waves and the king fled to safety. His daughter was transformed into a mermaid who would spend eternity luring sailors to their deaths.

The Cantre'r Gwaelod story in the poem 'Boddi Maes Gwyddno' also blames a woman. It is Mererid who causes its submergence when she allows the well to overflow:

> Cursed be the maiden
> Who let it loose after the feast,
> The cup-bearer of the mighty sea.

This medieval version casts Mererid as an Eve who brings about man's banishment from the bountiful Eden of West Wales. In a later retelling, Mererid is so infatuated with a handsome guest at the feast, named Seithenyn, that she neglects her responsibilities, turning the tale into a warning about the dangers of lust.

By the seventeenth century, the story was updated with details about a protective dyke with gates that allowed for drainage of the sodden land at low tide. This was the same century in which the Dutchman Cornelius Vermuyden arrived in England, hired by King Charles I to drain the Yorkshire and Cambridgeshire fens using his innovative windmill-powered pumping technology. Seithenyn, the gatekeeper, was now in the firing line. In 1757, the antiquarian Lewis Morris suggested that he was too drunk to operate the gates: a lesson in the perils of intemperance.[8] Taking a different moral tack, the poem 'Clychau Cantre'r Gwaelod' by J. J. Williams accused Seithenyn of laziness.

> But through idle negligence
> Of the watch on the tower,
> The bells of Cantre'r Gwaelod
> Vanished beneath the water.

To my mind, it wasn't femininity, lust or drunkenness that sank the Lowland Hundred, but the arrogant king and his political inner circle resisting the truth about their situation and putting their faith in technology or gods, so that they could abdicate responsibility and continue life as normal. The idea of decadently feasting while a rising sea slaps the walls has parallels in our times, where so many cities are under threat of submergence, yet we continue to mine, pump and burn fossil fuels.

Developed countries are unwilling to give up their luxurious lifestyles or reduce their consumption, even in the face of over-whelming scientific data; even when we see with our own eyes the freakish weather events that occur, year after year; even when we watch footage of houses floating down flooded European streets, East Asian towns under water, flames rising from Siberian tundra and infernos in North America. Climate-change deniers argue that these events are caused by solar cycles, or that the data is faulty, or that scientists are part of a New World Order conspiracy to dupe and control the global population. Even some of those who don't deny the facts tell us that our technologies will address the problem, should the situation become critical, or shrug and say that humanity will always find a way to thrive. There are even those who say that global warming might be welcome, like those authors of a BBC webpage aimed at schoolchildren who listed several benefits, including healthier outdoor lifestyles, more tourism in the UK and easier access to oil in Alaska and Siberia.

I imagine Seithenyn as the conscientious scientific observer, alarmed by the evidence of climate change, frustrated by the public's complacency and the politicians' lack of action. He is helpless to do anything about the disturbing phenomena to which he bears witness but report to the authorities and wait for the inevitable. He drinks too much alcohol, not because he is morally decrepit, or lacking discipline, but to self-medicate for his chronic anxiety. This is not uncommon in climate scientists today. Diana Six, an entomologist who works in Montana's Glacier National Park, claims that her daily observations of accelerating ice melt have led to a 'constant, haunting depression'.[9] She says, 'Somewhere along the way, I had gone from being an ecologist to a coroner. I am no longer documenting life. I'm describing loss, decline, death.' In 2015, climate activist Lise Van Susteren told *Esquire* magazine, 'So many of us are exhibiting all the signs and symptoms of post-traumatic disorder – the anger, the panic, the obsessive intrusive thoughts.'[10]

The well-keeper Mererid represents the workers in low-income countries at ground zero of climate change. They are not the main cause of heat-trapping emissions, and enjoy few of the luxuries that industrialisation provides, but are most at threat from its consequences. These include the Madagascans suffering a drought that has left 14,000 people in 'catastrophic conditions',[11] the Bangladeshis displaced by the worst South Asian flooding in living memory,[12] and the Alaskan tribes fleeing their homelands as rivers burst their banks and villages sink into melting permafrost bogs.[13] There are striking parallels between the Lowland Hundred myth and the situation in Guyana, one of the poorest countries in South America. For centuries its capital Georgetown, on the north-east coast, has been protected from the Atlantic Ocean by a 450-kilometre wall, built by the Dutch when they settled on reclaimed marshland. Today, most of its population live below sea level. The ocean regularly spills over the top, while sea water contaminates their wells. But rather than facing up to this reality, Guyana has bet its future on oil after Exxon discovered reserves beneath their seabed in 2016. An Exxon executive said it was akin to a 'fairy tale'. But with increased emissions contributing to higher sea levels, the result for Georgetown will be more akin to the fate that befell the Lowland Hundred.

In Cardigan Bay, 16 kilometres north of the drowned forest on the beach of Borth, the village of Fairbourne is facing this same grim prospect. Fairbourne was built on saltmarsh in the 1880s as a site for brick and tile works, with its own light-gauge railway station. In 1940, trapezoid concrete blocks lined the beach to defend against tanks should Hitler's German army invade. In the 1960s and 1970s it became a popular holiday spot but soon there was a new threat of invasion; this time from the sea. Many of Fairbourne's 420 properties are already 1.5 metres beneath sea level. Violent storm waves often crash over the anti-tank defences, breaching the shingle bank and concrete wall. Even when the sea is at bay, groundwater levels are so high that roads and gardens flood. The threat comes not only from the sea but from the river estuary to its northern side and rain run-off from the steep, quarried hills behind it. The risk of flooding is so great that life in the village will no longer be feasible after 2054.[14] The government has refused to fund further sea defences after that and recommends a managed retreat, which will turn the inhabitants into Britain's first climate-change refugees. Property prices have crashed. Mortgages have been denied. Insurance premiums have skyrocketed. Many of the residents are angry, claiming that the rate of sea rise is not a given, and that proper investment could save the village for their children to live in. Others are fearful that the place they love could return to salt-marsh within a few decades.

Visiting Fairbourne in 2021 I caught a glimpse of what it might be like in the days before the final flood. Wales was under Covid-19 restrictions, so the village looked deserted. No cafes were open. The car park was empty. The light-gauge train tracks lay silent, station gates closed. Empty plastic crates were piled up outside the minimart. One dog walker braved the shingle in an icy wind, a gull circling above him. Outside the train station, the wall of a public toilet was painted with a mural of the beach, the concrete teeth of its sea wall guarded by a merman with a silver-scaled tail, clutching a trident. With his noble visage, bushy beard and outstretched arm, pointing downwards, forefinger extended, he reminded me of God in Michelangelo's painting *The Creation of Adam*. Except in this case there was no Adam reaching back to accept the spark of life from his creator. Humans were gone. All that remained was concrete and stone beneath a cloudy sky.

I wondered if this merman was connected to the legendary Welsh figure Dylan, who was half man, half fish, and purportedly lived further up this same coast. North of Cardigan Bay, towards the town of Caernarvon, an enormous glacial erratic sits on the shingle beach, its triangular form mirroring the shape of the hills on the peninsula behind it. People have been gazing at this boulder since it was left by a moving wall of ice 12,000 years ago, wondering why it was there and what it might mean. On its flank is carved 'Maen Dylan', or Dylan's Rock, the name bestowed on it in *The Mabinogion*, a compilation of stories from the twelfth and thirteenth centuries, based on preceding oral traditions with uncertain chronological origins. About 900 metres out to sea, near the rock, is an oval reef. At extremely low tides, the tops of

the stones emerge, like turrets of a sunken castle. Local legend has it that this is Caer Arianrhod, the palace of Princess Arianrhod.

The story goes that her uncle, Math fab Mathonwy, was told by the gods that he must always keep his feet placed on the lap of a virgin when he wasn't at war, or he would die. Duly, virgins were supplied to keep his feet aloft. However, after the passing of one of these virgins, it was suggested – by her brother of all people – that Arianrhod should become the next foot rest for the king. But when put to the test, she turned out not to be a virgin at all. Immediately, she gave birth to a son, Dylan. She watched in horror as he writhed on the ground, crying his little lungs out. Ashamed and frightened for her life, Arianrhod fled. As she did so, she dropped behind her a lump of flesh, which her brother stashed in a chest. That quivering blob grew into a warrior king named Lleu Llaw Gyffes, whose fortress was linked to the Iron Age hill fort on a mound of earth further up the beach, another feature created by a retreating glacier. The remaining son, Dylan, was baptised, at which point he plunged into the sea, and became transformed into a human–fish hybrid. His gravestone is said to be the strange rock on the beach. The name Dylan is a combination of the Welsh prefix *dy-* ('toward') and *llanw*, meaning 'tide', 'flow' or 'flood'; in *The Mabinogion* his full name is Dylan ail Don, which translates as 'Dylan the Second Wave'. Some believe that he might have been based on an extant local sea god, enshrined in Celtic oral lore during the Iron Age, which then became linked to founding Welsh myths before it entered written texts in the twelfth century. It might therefore be that the depiction of the merman on the toilet wall in Fairbourne was a plea to those old gods, from the era of the Mesolithic floods, asking for protection during this second great rising of the seas.

As I departed Fairbourne on a hilly coastal road, I glanced back down at the village in the marsh, streaked with drainage channels, waves crashing on the shore and the estuary glistening behind it. It looked small and vulnerable, glued to its aquatic fate like a moth in a toilet bowl. Perhaps I would return here as an old man and look on its ruins, silted and bladderwracked at low tide, crabs scuttling on the railway tracks, reeds pushing through ruins. I might walk on its sandy bones as I did on the tree roots of Borth. Then in five hundred years' time they might tell morality tales about the 'Fall of Fairbourne' and the folly of trying to resist a rising sea.

—

I travelled inland from the west coast, through the winding mountain roads of the Snowdonia National Park in search of another sunken kingdom, Lake Bala, known in Welsh as Llyn Tegid. This ribbon-shaped feature, 6 kilometres long, was formed in the Ice Age when a glacier pushed through a geological fault line, forging a valley. When the ice retreated, terminal moraine blocked the water drainage, creating the largest natural lake in Wales. Just like Loch Ness in Scotland, also formed by a glacier in a geological fault line, this chillingly deep body of water exerts a dark imaginal power. In 1909 the folklorist Marie Trevelyan described how two men in a boat tried to ascertain the lake's depth with a plumb line. As the weight descended, a voice boomed, 'Line cannot fathom me. Go, or I will swallow you up!' Terrified, they fled. Trevelyan described how a brave diver also attempted to measure its depths but was shocked to spot a dragon coiled asleep at the bottom of the lake. It was the first sign that Llyn Tegid, like Loch Ness, had its own monster. A decade after Trevelyan's book was published, a dinosaur-like beast was glimpsed by a local and subsequently named Teggie.

I parked at the light-gauge railway station by the village of Llangywer and entered the lake's south-eastern shoreside via a wooden gate. Its littoral zone was a mesh of stones and washed-up reed fragments, scattered with sun-bleached driftwood, like the bones of Teggie's victims. On the skyline a swell of wooded hills was dazzled by shards of sunlight. The waves were choppy and a miniature surf lapped the pebbles. But in the shallows the water was still and crystal clear, burst through with clumps of reed. Behind me, the rise and fall of the lake's levels had washed away soil from beneath the roots of oak trees, leaving them on spindly legs, furred with moss, like Triffids stalking the shore-line for prey. Some seemed to step tentatively into the water. A few were toppled into the waves, branches like flailing arms, not waving but drowning. There was an uneasy sense that something calamitous had happened here. But this was undoubtedly the effect of folklore on my imagination, for I knew very well that this was another Welsh lake that concealed a sunken town.

There are two legends that tell of how the original Bala was lost. In the first, the town was the home of Prince Tegid Foel, a wealthy man prone to acts of cruelty who surrounded himself with equally fiendish friends. One day, he threw a banquet in his palace, plying guests with wine and entertaining them with a harp player. As he performed, the harpist heard a voice in his ear, whispering, 'Vengeance will come!' But when he looked around, there was nobody there. This happened again and again. The more it continued, the more the harpist became convinced that this was no auditory hallucination. Something terrible was about to happen. During the interval, he fled the palace and ran to higher ground, where he rested a while to gather his wits. Soon, a chorus of sweet birdsong lulled him into a deep sleep. When he awoke, he blinked his eyes in confusion. Bala was gone and before him glistened an enormous lake. The town and all its people had been drowned.

In the second version, Bala contained a well, which was guarded by a spirit. She appointed a guardian to place a lid on the well every evening to stop devils flying out from the underworld and flooding them all to hell. But after a night of drunken revelry, her employee forgot to carry out his duties. The demons were released and the well overflowed, filling the valley to create a lake. This version persisted through the centuries and by the 1700s, many believed that it had happened within recent history. In his 1901 book *Celtic Folklore: Welsh and Manx*, John Rhys, a professor at Oxford University, spoke to an array of people, from farmers to academics, collecting their recollections and stories. He was careful to clarify where these accounts originated, and how they were passed down through the generations. In one instance, he talked to a man named William Davis, who had found notes by the poet David Jones written in the mid-eighteenth century. In 1735 Jones met a Bala man who told him that elderly folk had memories of a water well named Tfynnon Gywer, at the centre of the original town. It was 'obligatory to place a lid on the well every night,' he explained to Jones. 'But one night it was forgotten, and by the morning, behold the town had subsided and the lake became 5 kilometres long and 1.6 kilometres wide. They say, moreover, that on clear days some people see the chimneys of the houses. It is since then that the town was built at the lower end of the lake.'

This might seem like a fairy tale, but there are parts of the world where this is really happening. In Kenya, the great lakes – Bogoria, Naivasha, Nakuru and Victoria – are flooding thanks to shifts in tectonic plates in the Great Rift valley, combined with record rainfall and run-off caused by unsustainable farming practices. Hundreds of thousands of people have been displaced. Lake Baringo expanded by 2 kilometres in just three years, its waters gushing through the town of Marigat, destroying crop fields, submerging buildings, unleashing hippos and crocodiles into the streets. In 2022, journalist Carey Baraka travelled to the shores of Lake Turkana where the El Molo people have had to move their village to escape the waters. 'Maybe there's a broken spring in the ground, and the rocks have cracked, so it's throwing water up,' the elders told Baraka.[15]

In the west of Ireland, Lough Funshinagh flooded in 2016, expanding to twice its size. It has remained that way ever since. The lake used to be drained naturally by a 'swallow hole', which worked like a bath plug, the water spinning away in a whirlpool once a year, leaving behind a swampy wetland of reed beds and rushes that harboured nutritious tadpoles, water beetles and freshwater plants for swans and geese to eat. But in the permanently flooded lake, the reeds are submerged, the seasonal cycles have been disrupted and the birds no longer come. The surrounding farmland is under water and homes are piled up with sandbags. 'The sad thing is nobody knows what happened,' one local said. 'Whether it is because of climate change or a collapse in the underground caverns or some obstruction that has got in the way, nobody knows.'[16] It is not hard to imagine how traumas like these might resonate through these communities' storytelling for centuries to come.

If there ever had been a human settlement in Llyn Tegid, perhaps far back in the Palaeolithic when the great ice wall dominated the northern skyline, or during the centuries-long drought of the Bronze Age, it might have been lost in a similar way to these farms and villages in Ireland and Kenya. Nobody would have been able to explain how and why it happened. Perhaps ripples from such distant events inspired Welsh tales many thousands of years later. It is impossible to say for sure. The version of the Lake Bala myth we know today was primarily shaped through inventive retellings by imaginative folklorists throughout the 1800s and early 1900s. But even those were driven by environmental anxieties. In 1798, the naturalist cleric William Bingley described how 'overflowings' of the lake would happen 'when the winds rush from the mountains, when they drive the waters before them . . . rising in stormy weather very suddenly from the joint force of the winds and mountain torrents, sometimes 8 or 9 feet in perpendicular height and almost threatening the town with destruction'.

Standing on the banks of Llyn Tegid, I could see that those forces remain a threat today. A sign bolted to a wooden fence warned: 'Please do not throw the rocks or stones into the lake: they form part of the bank protection during floods and storms.' Even an innocently skimmed stone could lead this fragile environment one tiny step closer to permanent submergence. Indeed, it has been foretold that the lake will one day rise again and flood not only the modern village of Bala, but its neighbour, Llanfor. According to a popular eighteenth-century rhyme:

Bala old the lake has had, and Bala new
The lake will have, and Llanfor too.

Local lore says that if you stand on the lakeside, you might still occasionally hear the voice that warned the harpist at Prince Tegid Foel's feast, 'Vengeance will come, vengeance will come!' Then you will hear this reply: 'When will it come?' At which point, the spectral voice will answer, 'In the third generation, in the third generation!' Future generations would be punished for the sins of their ancestors, in that same way as it is our descendants who will suffer the environmental consequences of our actions – or inaction – today.

Ancient prophecies haunt not only Bala but the riverside town of Carmarthen in the south-west of Wales, 14 kilometres along the Towy estuary. Its people live under a flood warning from the wizard Merlin, who was said to have been born there. Famous as the mentor to King Arthur in the twelfth-century stories by Geoffrey of Monmouth, Merlin was based on figures from Celtish mythology and the legendary sixth-century figure Myrddin the Wild, who went mad after a battle and fled to the woods, where he became a prophet. One of Merlin's predictions was that his hometown would be fatally flooded:

Llanllwch has been,
Carmarthen shall sink,
Abergwili shall stand
Carmarthen, thou shalt have a cold morning,
Earth shall swallow thee, water into thy place.

In folklore, the fate of Carmarthen's populace rested on an ancient oak tree which stood to the north-east of the town, near a bend in the Towy River. The prophecy read:

When Merlin's Tree shall tumble down
Then shall fall Carmarthen town!

In the 1800s, Merlin's Oak was a popular meeting point, and the site of festivals. The noise and disruption so incensed one of the street's shopkeepers that he poisoned the tree. By the early twentieth century, the oak was a skeletal husk, protected by iron railings on the junction of Old Oak Lane and Priory Street. At the end of the 1970s it was set on fire one night, rendering it little more than a charred, upright log. These pitiful remains were removed from the junction to improve traffic flow and that was the end of Carmarthen's protection. The Towy broke its banks and flooded the town in 1981, then again in 1985, prompting major improvements in its defences. But those new works were no match for the great storm of October 1987. The flood happened so quickly that many inhabitants became trapped in their homes as parts of the city were transformed into a lake. Stricken people and their pets were rescued from upper windows by rescuers in boats. In nearby Llandeilo, a railway bridge collapsed and a train plunged into the water below, killing four passengers.

The junction on which Merlin's oak stood is now a mini roundabout. On the pavement beside it, a new tree has been planted in a wooden pot. It's a noble effort to appease the legendary prophecy but the town remains troubled. In 2018, Storm Callum burst the seams of the Towy flood-defence wall, sending residents fleeing their homes, washing away a flock of terrified sheep and leaving a pub under water. 'The volume of water was simply too much for the flood defences,' said council member Stephen Pilliner.[17] By 2021, the repairs had still not been carried out and the river flooded another three times in a year. Authorities blamed storms and Covid for the delay, while local businesses complained that they were carrying out precautionary evacuations every few months. The Secretary of State for Wales told local councillors, 'We all agree that flooding is a result of climate change and its impact on flood management, and that solving this will take considerable research and substantial sums of money.' The question is whether short-term fixes can ever be enough, bearing in mind the scale of change that is happening. Huwel Manley from Natural Resources Wales said, 'We will have to manage expectations on how much flooding can realistically be prevented.'[18]

Reading about these events it's hard to shake the image of the prophet Merlin, staff in hand, eyes wild, pointing a tremulous finger towards the future, warning us that despite our engineering ingenuity there is no escaping nature's wrath. It's hard not to hear whispers of 'Vengeance will come, vengeance will come!' as the oil industry initiates new drilling projects in fragile wetlands while the public is forced to pay the cost of building walls, fighting wildfires and shoring up sinking towns. It's difficult not to think of the complacent King Gwyddno or the cruel Prince Tegid Foel and his sycophantic courtiers when billionaire despots live decadently in their protected citadels as waters flood their realms.

Admittedly, such imaginings have limited use. These myths don't tell us what to do in a time of climate crisis. Whatever disasters inspired them originally, we can never know what happened, or how the victims coped. We know only that they were powerless to stop the encroaching waters. Perhaps the Victorian versions of sunken-city legends do more harm than good, blaming individual error and moral lassitude for the catastrophes in the Lowland Hundred, Lake Bala and Ys. It's a narrative that suits big business and oil corporations – where the onus is on us to change our wicked ways and eat less, drink less and do more recycling, rather than on the rulers of this world to change the environmentally destructive economic and technological system in which we have no choice but to live. But what's great about mythology is that it's ours to adapt and retell. From a twenty-first-century perspective, the question asked by these legends might be: when the evidence shows that the water is rising, why continue to build in the same way as before? Why live in such defiance? Perhaps the new lesson should be that we must not rely on better walls, higher walls and more walls but adapt our way of life to seasonal flood and rising oceans. Our long-term survival relies not on the temporary fix of bolstering sea defences after every flood, but on stopping development on floodplains altogether, returning estuaries and wetlands to their natural state of flux and reorganising our society on higher ground in a sustainable way that relies less on fossil fuel, resource consumption and perpetual economic growth. It might not save us in the end, but it will mitigate against the worst and help bestow a habitable biosphere on those who inherit what's left after the flood.

—

On the north coast of Wales, there is another lost kingdom between the island of Anglesey, off the western mainland, and the Wirral in England to the east. Once a forest of oak, alder, birch and hazel grew where the Irish sea pounds against a limestone promontory known as the Great Orme. In 1893 the writer Edward W. Cox walked on the Great Orme towards Llandudno with a farmer, who pointed at the sea and explained it was the site of Helig, a land that stretched all the way to Puffin Island. When Cox objected that the water was too deep for that to be possible, the farmer replied, 'You may know better than I, but I was told it as a boy, and all the people hereabouts believe it.'

Like the Lowland Hundred, Tyno Helig was purported to exist in the fifth and sixth centuries, ruled by a king named Helig ap Glannawg, until it was destroyed in a flood. One version goes that Helig's daughter Gwendud fell in love with a man named Tathal. He might have been the son of a local baron but he was not a noble, and therefore unable to marry a princess. But the besotted couple were desperate to be together and when they heard about the imminent release of a captured Scottish chieftain, they hatched a plan. Tathal offered to escort the chieftain home to Scotland but killed him before they reached the border and blamed bandits for the death. He returned bearing the deceased man's golden torque – a symbol of nobility, and his ticket to a royal marriage. During the wedding reception, however, the Scottish chieftain's ghost turned up to swear vengeance on the couple. It would not be the couple themselves who suffered, he warned, but their great-grandchildren. So it came to pass, many decades later, during a feast in which four generations of the family came together in celebration, that a crack opened in the cellar and water filled the palace, killing everybody but for the grief-stricken octogenarians, who fled to higher ground to watch everything they loved destroyed.*

The story attached itself to the strangely angular rock formations that could be seen at super-low tides near the coastal town of Penmaenmawr in Conwy Bay. In 1864, Charlton Hall journeyed to the rocks on a boat and observed what he described as the remains of a grand old hall, covered in seaweed. Forty-four years later, William Ashton saw 'three sides of a large square, with a large rectangular recess at the south-west side . . . well defined

---

* This same story has been attached to a number of flooded lake legends in Wales. This has led some, including F. J. North, author of *Sunken Cities* (1956), to believe that the myth of Tyno Helig might have been imported to its north coastal location in medieval times, and later embellished by eighteenth- and nineteenth-century folklorists.

by straight and almost continuous lines of wall'.* He concluded it was 'quite impossible for anyone to view these 350 or more yards of strictly rectangular remains and to entertain the slightest doubt as to their having been human handiwork'. Most of these features have since been identified as sea-sculpted glacial deposits† but the legend had taken on a life of its own, regardless. And there was more to stimulate the folkloric imagination. To the west of Penmaenmawr was the Menai Strait, a waterway between Anglesey and the mainland. At low tide, large flats known as the Lavan Sands emerged, where sightings of tree stumps were reported in the nineteenth century, along with a causeway leading towards Helig's ruined palace. Over by the Great Orme was Maen Rhys, a rock visible in low water, said to be the place were a shepherd named Rhys sat to watch his flock graze on fields that now lay beneath the sea. This might have been possible when sea levels were lower, but it would have been long before the fifth century, back in the time when a forest flourished along the now submerged coastline. Glimpses of it could occasionally be seen at low tide after stormy weather. For instance, to the east of the Great Orme, on the beach of Rhyl, where archaeologists discovered a 6,550-year-old antler mattock, an axe, a fish trap and tree stumps in slabs of peat.

I had seen photos online of these stumps, and wanted to find one for myself, so I made Rhyl my final stop in Wales on a cold afternoon. The town's heyday was in the early twentieth century, when visitors flocked there on holiday from their homes in smoggy industrial cities and mining towns. Like many British seaside resorts, it suffered from the effects of cheap flights and

---

* Described later in his 1920 book *The Evolution of a Coastline*.
† There is speculation that there was a fish weir in this location in the sixth century, which might explain a few of the anomalies seen over the years.

package holidays. On a weekday after rain, the place seemed as deserted as the village of Fairbourne. Amusement arcades were open, lights blazing, but I could see nobody inside. However, after I parked on the seafront, I was reassured to hear the droll tones of a bingo caller drift in the breeze as a Liverpudlian couple argued about how much parking they needed to pay for. 'We're not going to be here *that* long!' said the woman with faint disgust.

I headed onto an expanse of ribbed sand, stretching for 3 kilometres from the Clwyd estuary to a zone called Splash Point. It was low tide and the sea was a kilometre out. I wasn't sure where the remains of tree stumps had been spotted, so I walked forth with no aim. The sand was coated in a watery film. My feet squelched into pockets of quicksand, engulfing my boots, making the walk heavy going. The only landmark between me and the distant surf was a wooden beacon on a mound and my only company lugworm fishermen, scattered like Gormley statues in waterproofs. I paused to catch my breath and gazed towards the horizon, trying to imagine the forest that once grew there. The skyline bristled with wind turbines, where once there were trees. Viewed at certain angles, the sails of multiple turbines became superimposed on each other, so they looked like Da Vinci's Vitruvian Man, and just as beautiful in their graceful proportions, arms spinning for the salvation of a stricken planet. Beyond them, I could see the oil rigs of the Douglas Complex, structures that contribute to the global warming that threatens to put the town of Rhyl completely underwater by 2050 if sea-level rises continue as predicted.

The oil industry has been aware of the consequences of its actions since the speech by the physicist Edward Teller to a symposium of executives, economists and scientists in 1959. He told them:

Carbon dioxide has a strange property. It transmits visible light but it absorbs the infrared radiation which is emitted from the earth. Its presence in the atmosphere causes a greenhouse effect [...] It has been calculated that a temperature rise corresponding to a 10 per cent increase in carbon dioxide will be sufficient to melt the icecap and submerge New York. All the coastal cities would be covered, and since a considerable percentage of the human race lives in coastal regions, I think that this chemical contamination is more serious than most people tend to believe.[19]

Rather than act upon this, ExxonMobil (then named Humble Oil) paid for a newspaper ad in 1962 that boasted, 'Each day Humble supplies enough energy to melt 7 million tons of glacier!'[20] In 1968, the Stanford Research Institute's report to the American Petroleum Institute warned, 'If the Earth's temperature increases significantly, a number of events might be expected to occur including the melting of the Antarctic ice cap, a rise in sea levels, warming of the oceans.' Another internal report made by Exxon's own scientific researchers in 1982 predicted that the global warming caused by its products could lead to catastrophic sea-level rise and flooding.[21]

Despite the clear evidence provided to them at their own behest, the world's major oil companies have been accused of spending decades either ignoring global warming or funding contrarian research projects to contradict it. They have promoted climate-change misinformation through advertising, lobbying and press propaganda to keep the populace ignorant of the threat, so that they can continue to make money for themselves and their shareholders. When forced to address global warming by growing public and political concern over the scientific data, they introduced the concept of a 'carbon footprint' to shift the burden of responsibility onto the individual, rather than corporations or governments, suggesting that it was our lifestyle choices that were really to blame. We were the consumers and they were dutifully servicing our insatiable demands. It was up to us to change our ways. But, in reality, theirs is the crime that will result in the floods that curse the next generation and send populations fleeing deluged coastal towns, not the decadence of everyday people.

I had come to find ancient tree stumps, but the modern steel forest out to sea gripped my imagination, observed as it was from a coastal town that might one day drown like the kingdom of Tyno Helig. In that moment I perceived the mythic potential of the present where an epic battle for the planet raged in the waves between the proponents of renewable energy and the oil barons, desperate to cling onto their wealth and power. After floods and fires ravaged the great civilisation, the untended wind turbines would tumble over and join the shattered skeletons of rigs in a thick slick of oil, tangled with the corpses of sea birds. This toxic morass would be swathed in silt, rotor blades jutting up like tombstones, ribboned with kelp. These ruins would lie beneath the swollen Irish Sea, many kilometres out from the shrunken coastline. Stories of the ancient battle would outlive the memory of the events themselves, linked in the local folklore to those lumps of limpet-encrusted alloy metal exposed on the seabed after storms. The names Exxon and Shell would be whispered around campfires instead of Gwyddno and Helig; Greenpeace and Extinction Rebellion instead of Seithenyn and Mererid. In hushed tones people would tell of old kings who unleashed the sun energy trapped beneath Earth's crust so that they could attain the power of gods, only to doom themselves to an inexorably rising heat that melted all the world's ice and flooded their kingdom forever.

The afternoon darkened like my mood as I trudged across the monotonous sandflat towards a herd of mechanical diggers on the beach at Splash Point, where men in hi-vis suits were constructing a sea defence. The air vibrated with clanks and chugging. It was a desolate scene: machines digging the beach while other machines in the distance drilled for oil to fuel those digging machines. And, all the while, spinning wind turbines out at sea tried to make amends for the sorry affair. By the time I reached the construction site I had lost my enthusiasm for finding a tree stump. Which was fine, because there was no drowned forest in Rhyl that I could see that day. But what difference did it make if I didn't observe a fragment of dead tree? The forest had been there all the same, somewhere beneath the sand. It was like dark matter in the universe, undetectable to human senses and yet mysteriously holding galaxies together, or the phenomenon of global warming, which is invisible, yet shapes our reality in a profound way.

This is the eerie aspect of sunken lands. They are places we cannot always see. Events we cannot always remember. Facts that might be fictional, and fictions that might be factual. But despite their elusiveness, they exert a strong gravitational pull on us. Their uncertain features emerge to haunt us, briefly, when the moon draws back the tide to reveal a rock, reef or tree stump – fragments revealed after a storm has shorn away the sand, before they are dissolved, washed away, or covered over again. They intimate an epoch of extreme climate change that mirrors our own, shadowing us with their ruins, enticing our minds to put the broken pieces back together and view the whole picture. We can never truly know the origin of a flood legend or see what once lay beneath the water. To tell the story of sunken lands, we must deduce what is absent and fill in the gaps. In doing so, we collaborate in a folklore that is constantly unfolding, preserving ancient anxieties while expressing our own in the present, so that they might carry forward into the future, there to inform our descendants yet to be born.

That evening I slunk wearily into my room in St George's Hotel in Llandudno, a coastal town on the Great Orme, 24 kilometres west of Rhyl, where I rested on the bed, watching *Tales of the Unexpected* on TV. By the time I felt hungry enough to go out for dinner, it was dark. I stepped onto the balcony and looked beyond the promenade to the Irish Sea, where I was surprised to see the sparkling lights of a distant coastline. I struggled to work out what I was observing. Perhaps it was the coastline of the Wirral, in England, but that was too far away and in the wrong direction. I whipped out my phone and checked Google Maps to make sure I was not being stupid. No, nothing was supposed to be there. Only sea. But then again, clearly there was *something* there, where the legendary lost kingdom was said to have existed. Praise be! I thought. This was the land of Tyno Helig, arisen from its watery grave to offer me a vision of North Wales' mythopoeic landscape.

It quickly dawned on me that I was looking at those windfarms and oil rigs I'd seen earlier from the beach in Rhyl, lit up to warn away boats and planes. But the magic was undiminished. For while I had not found the sunken forest, nor Helig's palace, I had momentarily crossed into the ghost world of the collective cultural memory, where the narrative transmission signals from centuries of storytelling, inspired by ancient climate disasters, warned me that the tale is never truly over.

4

THE SHRUNKEN FEN

During their occupation of East Anglia, the Romans inflicted brutal punishments on the Iceni, a Celtish tribe, chopping off their ears and noses, hunting them through the woods with dogs, then whipping them to death. One day, they seized Rowena, the daughter of an Iceni priest named Mandru, and imprisoned her in the palace of the Roman governor, Valerian.

A furious Mandru gathered his fellow priests and tribe leaders at the temple of their sea god to share their accounts of enslavement, torture and murder at the hands of the colonisers. A troop of Roman soldiers burst in, hacking some unfortunate rebels to death and dragging others away for crucifixion. Only Mandru escaped into the woodlands, where he made plans for his revenge.

Months later, a mysterious bearded traveller appeared. Whenever he encountered a British slave, he whispered in their ear, 'Friend, arise this night and be gone. Destruction comes fast upon this city and if thou tarry, there shall be no escape.' Most took him at his word and packed up their belongings, but others laughed him off as a madman.

That night, while the Roman sentinels were asleep, the believers fled to the woods, encountering other Britons from neighbouring settlements, all of whom had received the same warning from the same stranger. In that instant, the man himself appeared, now clean shaven. They recognised him as their exiled leader, Mandru. 'The gods are angry with Rome,' he explained, 'and they will destroy their towns and cities tomorrow.'

The next morning, the Romans woke up to the sound of a ferocious gale howling through the woodlands. They were shocked to find that their Iceni slaves had departed. As they checked the empty huts, they noticed something astonishing rise on the skyline, where the North Sea crashed. It was a mountain of water, 90 metres tall, moving at high speed. The Romans were helpless as it smashed through their fortifications and houses, then crashed through the trees towards the low hills where the Iceni huddled in fear.

When the storm abated, the forest had been replaced by an inland sea, dotted with islands. Treetops poking above the surface snagged the cloaks of drowned Roman soldiers spiralling on the current. Mandru turned to the Iceni survivors and announced, 'These parts shall never more become forest. The sea, our great deliverer, shall always be present here, in token whereof it has been decreed that we shall be known henceforth as Gyrvii or marshmen, in place of Iceni, the slaves of the Romans.'

From that day, the marshmen learned how to hunt and fish in the watery land bequeathed to them. When the Romans returned, they could never control the fens the way they had previously. Their dykes and roads did not give them dominion over the canny fen dwellers, who understood how to thrive in the treacherous boggy quagmire.[1]

As I swooped into the Fens, north of Cambridge, Hawkwind blaring from my car stereo, it didn't look like a disaster zone. Not at first. Only a green and yellow flatland unfolding towards a limitless horizon beneath an eternally shifting sky. But the more I traversed its long, straight roads and water channels, criss-crossed with concrete bridges, the more I sensed I was on an artificial construction of interconnected causeways, dykes and embankments: an East Anglian version of the Lowland Hundred. Like the ill-fated Welsh kingdom, its topography had been engineered for the purposes of water control: dammed, gated and sluiced, drained and ploughed, seeded with wheat, barley and sugar-beet, then harvested for profit. But despite all this ingenuity and innovation, the land was sinking below sea level, year after year, haunting its human colonisers with the spectre of an imminent and calamitous flood.

In a chug of razorblade guitar riffs and spiralling synthesiser bleeps, I drove past great expanses of soil ploughed into deep grooves, patrolled by buzzing electricity pylons. Grassed embankments devoid of flowers. Waterways empty of wildlife, surfaces like glass, reflecting bruised clouds. Fallow fields covered by flapping sheets of polythene, weighed down with stones. Farm buildings slung together with telegraph wires. Houses with wooden-jetty driveways over streams, like shacks on a Louisiana delta. Ruined windmills consumed by ivy, towered over by gleaming wind turbines. There was beauty to the sharp geometry, block colours and epic proportions but also sadness at what had been lost and what could have been. For this was the site of what historian, author and environmental campaigner Ian D. Rotherham describes as an 'ecological catastrophe'.[2]

The story begins at the end of the Ice Age, when a dense oak forest filled a lowland basin on the eastern rump of Britain, where the Fens are today, ranging out towards the hills, marshes and lagoons of Doggerland, the land that rolled east to Denmark and the Netherlands, grazed by bison, mammoth and aurochs. The gigantic, towering oaks supplied excellent wood for the Mesolithic and then Neolithic tribes, who used flint to cut and shape the material into shelters, boats and artworks. Some trees were so colossal and imposing on the landscape that they became sacred ceremonial sites, where wooden monuments were constructed.[3] Rivers brimming with fish wound their way through the forest and drained into a nascent North Sea, which was then little more than a lake.

As the great glaciers melted and the relieved land tilted downwards, the expanding sea encroached on Doggerland, shrinking it into a lowland corridor. The rivers became backed up in East Anglia's shallow basin and the forest regularly flooded, leaving silt deposits and widening pools of brackish water. The great oaks died and toppled into the mire, where they became preserved. Over thousands of years, rivers flowing into this basin dumped more and more sediment, turning it to swamp and marsh. As vegetation decayed in standing water it turned into peat, which piled up into raised bogs. Trees and shrubs colonised this new layer but inundations from the sea killed them too and covered them with clay and silt, on which sedge peat formed, broken through with rivers and streams.

Upon the decomposed bones of a lost forest world, another world emerged: a 300,000-acre wetland at the edge of the Wash, a bay in the North Sea, curving round the coastline from Lincolnshire in the north-west, through Norfolk to Cambridge in the south. In its prime, it consisted of marshes, lakes and rivers full of eels, frogs and leeches, abundant in wild fowl, otters, water voles and insects. Each spring it would flood. Then in summer it would dry out into fertile meadows and grazeable grasslands. The topography was in constant flux, not only seasonally but within greater cycles of fluctuation, where extended dry spells allowed people to settle and learn how to hunt, fish and clothe themselves in a transient world. The ebb and flow of floodwaters delivered nutrients to the land, creating environments for animals and plants to flourish.

But this coastal edgeland was not an easy place to live. Change was constant and its inhabitants teetered on the brink of catastrophic flooding. There are clues that the Bronze Age fen dwellers who faced the threat of inexorably rising waters attempted to appease the gods of nature. In Lincolnshire, swords have been exhumed in what would have been thresholds between the marshland and higher ground, thrown into the water for ceremonial or spiritual purposes. Archaeologists have considered that the intensification in this ritual at the end of the Bronze Age might be linked to environmental pressures at that time, including rising water tables, backed-up river systems and sea-level change.

Water levels reached their peak in the Iron Age, but when the Romans arrived in Britain the Fens were in a drier period. That didn't mean it was an easy place to conquer. Writing a few hundred years later, the Roman historian Dio Cassius describes how the rebellious Britons, led by Boudicca, fled to the wetlands where the centurions could not attack them, hoping to wear out the invaders. Undeterred, the Romans 'wandered into the pathless marshes and lost many of their own soldiers'.[4] This may have given rise to the folk tale that opened this chapter, about Mandru and his sea god's deliverance of a deadly tsunami. Once the Romans were established, they built the Fen Causeway, a raised gravel road that cut from west to east across the marshes, from Peterborough to Denver in Norfolk. They dug a ditch between the towns of Ramsey and Lincoln, known as the Car Dyke, and built forts along its length. This was the first ever attempt to drain the Fens, and the opening salvo in a 2,000-year-long human war against this mutable topography.

After the Romans came the Anglo-Saxons, who colonised its low promontories and islets. In the ninth century the Vikings ruled the Fens, and indeed the whole of East Anglia, under the Danelaw. When William the Conqueror invaded, they became the refuge of the outlaw Hereward the Wake. The Normans had stolen his family lands in south Lincolnshire and murdered his brother, leaving his head on a spike outside his house. To take his revenge, Hereward made his base in the Isle of Ely, surrounded by impenetrable marshes, and employed guerrilla tactics to confound and repel the enemy, hiding in the reeds and setting fire to willow and brushwood to burn the enemy's siege ramps, walkways and palisades.

In the centuries that followed, the Fens remained wild and lawless, hard to navigate, hostile for anyone not from those parts. Disorientating mists crept across the peatlands to the cries and whistles of unseen beasts. Stinking miasmas lingered over stagnant pools. Plumes of noxious methane, carbon dioxide and hydrogen sulphide burst through crusts of rotting vegetation. The nocturnal glows of their combustion were called 'lantern men', evil spirits who lured unsuspecting visitors to their deaths in the treacherous reed beds and bone-sucking bogs. A form of malaria known as marsh fever, or ague, was rife. While locals developed some natural resistance, many intruders and incomers fell prey to shivering, joint pain and a fitful death. To ward off the disease, fen dwellers would guzzle down copious amounts of brandy and opium, extracted from locally grown poppies. The Fens became associated with moral, mental and physical decrepitude, loathed and feared by outsiders.*

There were repeated attempts to tame and drain the troublesome Fens throughout the Middle Ages. But the turning point

---

* As Rod Giblett has written in *Swamp Deaths* (2022), 'In the patriarchal Western cultural tradition, wetlands have been associated with death and disease, the monstrous and melancholic. If not the downright mad.'

came in 1620, when King James I decided to reclaim the land and extract its bounty for profit. He died before he could put his plan into operation, but King Charles I would later take up his father's mantle with gusto. In 1639 he hired a young Dutch engineer named Cornelius Vermuyden to carry out the ambitious scheme. Since the fifteenth century the Dutch had been using windmills as pumps to reclaim low-lying land for agriculture. Now Vermuyden did the same in the Fens, straightening and diverting rivers to dump their deposits into the Wash, using dykes and sluices to drain the fields.

These innovations were so famous that they began to feature in seventeenth-century retellings of ancient Celtish flood tales including those of Ys and the Lowland Hundred. They anachronistically incorporated this new Dutch technology to frame the moral message that whatever ambitions people might pursue for their own ends, using ditches, walls and sluice gates to achieve dominion over the world, they could be undone by the wrath of gods. The storytellers might have been influenced by the repeated failure of Vermuyden to conquer the Fens. For while his windmills successfully extracted the water from the drenched bogs, there was a price to pay. As the peat dried, it shrank, lowering the level of the land beneath that of the surrounding water channels, leaving it even more vulnerable to flooding each spring. There ensued a desperate battle between the drainers and the forces of nature as reclaimed land was flooded and then reclaimed again many times over.

Locals abhorred the project. For centuries they had lived off the land, trapping eels, reed cutting and hunting wild fowl. Now their very livelihoods were at risk as these intruders sucked dry their wetlands and sewed them with plants and flowers for export to the growing empire, and to fill the coffers of corporations and kings. They feared that they would eventually be driven from their homes and forced to eke out a living in unfamiliar territories where their unique set of skills was useless. A covert band of freedom fighters, the Fen Tigers, counterattacked by sabotaging equipment, damaging new ditches and setting fire to reed beds.

By this point, the subjugation of the Fens had become as ideological as it was financial. Drainage would rescue the fenland dwellers from their lowly state, improving their health and morals to that expected of good English Christians. God had bestowed on humans the earth as their garden after the Great Flood, commanding Noah, 'Be fruitful and multiply and fill the earth and subdue it, and have dominion over the fish of the sea and over the birds of the heavens and over every living thing that moves on the earth.' Therefore, the resistance of the fenlanders was an act against the divine order. As the historian James Boyce has written, 'The people of the Fens were depicted in the same way as indigenous people defending their homes everywhere – idle beings who had forfeited their moral right to possess their country because they did not farm it as God had ordained.'

This attitude did not change at the onset of the scientific age. Francis Bacon, the father of empiricism and a favourite of James I, wrote, 'The whole world works together in the service of man; and there is nothing from which he does not derive use and fruit.'[5] Nature was seen to work like a machine, made of component parts that could be observed, manipulated and controlled for the benefit of humans. And so the draining of the Fens in the Age of Reason was carried out in the name of 'improvement' with a righteous conviction as zealous as any religious ideology.

After the invention of the steam-powered pump in the nineteenth century, water could be kept at bay permanently, giving venture capitalists almost unlimited powers of reclamation. In 1850, engineers successfully drained Whittlesey Mere, a 10-kilometre-long lake, abundant in fish, wildfowl, reeds and sedge, which supplied surrounding villages with food, fuel and a means of income. Meanwhile, many embattled Fenlanders had developed laudanum addictions, thanks to large quantities of opium arriving on ships from India and China. This intensified the argument that drainage was not only economically vital but a means to the salvation of what was portrayed as an enfeebled, backward, drug-addled populace.

In the twentieth century, our world view became entrenched in what the late Mark Fisher described as 'capitalist realism': a perception that the near-universal system of private ownership, free markets and profiteering was the only one available and not really an ideology at all, but the *way things were*. This trapped us in an inescapable process of plunder and devastation, prioritising profit and economic growth over the ecosystem. By 2001, 99 per cent of the Fens had been drained, resulting in what Ian D. Rotherham calls 'the greatest single loss of wildlife habitat in Britain and maybe Europe'.[6]

The reason I was travelling through Cambridgeshire was to explore England's most sunken land, a place that used to be seasonally flooded, enriching it with spectacular biodiversity, but which had since been drained for monoculture farming, lowering the land further and further below sea level, leaving it dangerously at risk of cataclysmic floods as that sea level rises in our era of global warming. I had three days completely to myself, free of children and dog-care, with only Hawkwind for company. As my dented Peugeot barrelled across raised embankments, fields to either side, the band's lament to disrupted nature, 'We Took the Wrong Step Years Ago', chimed in my ears. By the time I approached the final fragment of original wetland in my silver machine, they were belting out 'Standing on the Edge', which describes humans in a ruined world using a scythe to cut the land, a spade to dig up its roots and a fire to burn the remnants, which are then blown away in the wind. This was an apt description of the fenland. A place that had been dug and pumped, drained of life, its peat beds left to dry in the sun. Two centimetres of soil are lost every year in a phenomenon known as 'Fen Blow', where surface dust is whipped up into a thick brown cloud that cloaks the sky in a dismal shroud, like a dystopian futurescape from *Blade Runner 2049*. But while this was today's grim reality, there existed a hole in time through which could be glimpsed the ancient fen, a portal to the flooded past that shows us where we went wrong and how things could have been different.

In 1899, entomologist J. C. Moberly bought two acres of undrained fenland for the handsome price of £10. He gave them to a new organisation, the National Trust, founded four years previously by Octavia Hill, Sir Robert Hunter and Canon Hardwicke Rawnsley, to offer the urban masses an escape from the grime and noise of their daily lives. Wicken Fen became their first nature reserve. It has since grown into 2,000 acres of reed and sedge fen, shallow lakes and scrubland. Water management techniques now keep them wet, rather than drain them.* Protective enclosures foster wildlife. Regular cutting maintains the reed-bed habitats. Free-roaming horses, ponies and cattle help spread seeds in hooves, coats and dung. Thanks to these measures it is the most species-rich area of Britain, featuring otters, newts, cuckoos and cranes, bitterns, marsh harriers and reed buntings, along with 5,000 insect species. This was as close as I could get to viewing what the naturally flooded land might have looked like before the Dutch engineers came.

---

* A project known as 'the Wicken Fen Vision' plans to broaden this to 53 square kilometres by the year 2099.

I parked on the gravel outside the public toilets and walked a short country lane to a modern visitors' building, which housed the ticket office and gift shop. On the lawn outside lay a bog oak: a tree from the sunken prehistoric forest, preserved for thousands of years in an airless tomb of peat, turned black by the reaction of tannins in the wood to soluble iron in the water. When the Fens were drained, these woodland zombies rose from the dead, only to be hacked into pieces and burned in fenland hearths. They still emerge from time to time, like the 13-metre-long trunk, older than Stonehenge, that was excavated from a farm in Wissington Fen in 2012 and turned into a banquet table for display in Rochester Cathedral. It shows how much our reverence for these trees from a lost world has grown in a time where our place on Earth has become suddenly tenuous and we face our own potential extinction. Like the sunken forest, all that remains of us might be fragments that emerge from the ooze. Gently, I placed a hand on the bark of the bog oak, just as I had on the tree trunks on the beaches of Pett Level and Borth. The thrill of touching organic objects from the Stone Age remained undiminished. It had become my ritual in the sunken lands. A means of tuning into ghost transmissions, resurrecting lost worlds, transmogrifying the ancient past into a resonant *now*.

Passing through the electronic doors of the gift shop, I stepped onto a raised duckboard walkway. Over the wooden fence, green sedge sprawled towards a skyline of scrub. I began to walk the elevated circuit, weaving deep into the fen. Tall grasses swayed in the breeze to the rhythm of my feet. Blossoming alder buckthorn twisted from mulchy swamp between enclaves of coppiced wood and black mirrors of standing water. I passed an original windmill used to drain the fens, made of lacquered wooden boards with brilliant white sails, restored and moved here for the historical education of visitors. The whole site had the feel of an open-air museum. Wooden arrows told me to walk anti-clockwise. There were playground areas and activity zones. Birdboxes and neat piles of logs. Copious tourist information boards. Beside a flooded claypit, full to the brim with reflected cumulus cloud, a sign told me about the history of the kiln that used to be there in 1880. Another board told me about brimstone butterflies, with wings that mimicked leaves. But in the shrubs beyond the photograph, I couldn't see a thing, possibly because they were disguised or possibly because they weren't there. There were information signs about birds such as willow warbler, chiffchaff and blackcap, in the locations they were most likely to be, describing the sound of their calls, with colour photos to show me what they looked like.

Should all these signs vanish, along with the elevated boardwalk, I'd be wandering thigh deep in ordure, without a clue what was supposed to be there among the reeds and rushes. It was as if I was walking through a description of a place, while the real place was concealed from my senses. Instead of the thing, I was accessing *data about the thing*. When I read descriptions of animals and plants, with their Latin names, I was viewing objects in the natural world through their classifications, as if they were merely bundles of properties, and the environment was simply a container in which they existed. This was the same conceptual prism through which those who drained the Fens viewed nature: as something separate, that existed only in relation to humans, there for our utility, whether for consumption or cultural improvement.

In *Dark Ecology*, Timothy Morton describes how this way of thinking originated during the traumatic and uncertain period of climate change at the end of the Ice Age, when the first farmers in Mesopotamia began to select crops that would be more palatable for humans, hardy and easy to grow, and then store them as surplus. The aim was to eliminate fear and anxiety 'by establishing thin rigid boundaries between human and non-human worlds and by reducing existence to sheer quantity'.[7] Our survival required the harnessing of the biosphere at any cost. Whatever made the land more productive for humans was prioritised. This gave birth to a 'logistics of agriculture' that today runs like a computer algorithm, all around the globe.

The paradox is that this drive to survive is also a death drive. It necessitates the continual felling, draining and crop selection that will ultimately break down the life-support systems on which we depend. It led to wetlands destroyed to make farmlands. Soils stripped of their nutrients by industrial-scale tilling. Rainforests cut down for cattle ranches and palm-oil plantations. Pandemics unleashed as species move into human territories with their parasites and pathogens. Insect species wiped out at the rate of 2.5 per cent each year by artificial fertilisers and insecticides. Entire food chains disrupted from the bottom up. The universality of our agricultural system leads us to assume that it is intrinsic to human behaviour, necessary for our survival and therefore unchangeable. But this is not an unavoidable reality. It's a behaviour pattern inherited from Mesopotamians at the end of the last Ice Age. 'We are still within this 12,000-year "present" moment, a scintilla of geological time,' writes Morton. 'What happened in Mesopotamia happens "now".'[8]

Our current solution to environmental collapse, however, is not to change our programme settings, but to cordon off bits of land to protect or re-establish wild environments so that we can continue as before. But this is the same pattern of thinking that caused the problem in the first place: that false impression of a boundary between us and nature; that there is such a thing as a solid, containable, comprehensible space that we can control. But, in truth, places are mysterious and uncanny, where the human and non-human are embroiled in complex interactions on infinite scales. There is no such thing as 'nature', argue dark ecologists, but a teeming multiplicity of entities, from atoms and blood vessels to stones and birds, to tin cans and flyovers, to radiation and wind, to oceans and stars. Humans are objects inside bigger objects, like the biosphere, the climate, and the solar system, with smaller objects within us, like bacteria, viruses, fungi and archaea. We are inextricably enmeshed. Nature is inside us, as much as we are inside *it*, whether we are in a protected fen or suburban cul-de-sac. If humanity were to become aware of this, we'd each feel every hurtful blow against the biosphere – every oil spill, decimated forest and plume of toxic smoke – as an attack on our own bodies. This shift in consciousness would make it hard for governments and industrialists to continue acting the way they do, for it would be evident to all that their actions were tantamount not only to ecocide, but potentially the genocide of the human race.

I stopped on a bridge and stared down at a raised waterway known as a lode. Some historians speculate that these were created by the Romans. So this might have been here for an extremely long time. I imagined that there might be things in the lode that wouldn't fit into the official fenland schema, such as brooches, coins, microplastics, oestrogen and brake fluid. They were concealed deep beneath an oily black film where pond skaters were busily engaged in a frantic meet-and-greet like coked-up speed daters. White clouds were mirrored glitchily in the concentric ripples that emanated from their twitching feet. Above and below, macro and micro, in constant interaction. There was so much more to this fen than I could comprehend, even with the help of an information board. Not only the colour of a butterfly, the drumming of a snipe, and the shape of an orchid, but their internal microbial worlds, their interactions with

other non-human creatures, their chemical and electric communication systems, their visual and auditory processes, maybe even their dreams and nightmares – phenomena way beyond that which I could access through my limited perceptions. This invisible dance of infinitely unknowable entities made the idea of owning a piece of land seem ridiculous and the notion of draining it monstrously irresponsible.

Towards the end of the duckboard circuit through Wicken Fen I arrived at a tiny piece of sedgeland that had never been drained: it rested on a peat layer that had been undisturbed for thousands of years, remaining elevated above the surrounding fenland. Without winter floods to keep it moist, the peat would quickly dry out and shrink, so it was being artificially inundated with the same technology that had once been used to drain the Fens. A windpump took calcium-rich water from the surrounding lodes and circulated it through this precarious relic. The guts of the pump gurgled and burped as the sails turned. It was as if I were standing by the bedside of a comatose patient on a life-support machine, hooked up to tubes and machines. On one hand, it felt pointless to hold on. But on the other, I could sense that this fragment of wetland was sentient and aware. Alive. Precious. Worth saving.

As I headed back to the National Trust centre, the walkway arcing through a swathe of sedge, I heard a noise like one of those Newton's Cradle office toys from the 1980s – *click click click click* – but I couldn't tell if it was a bird or an insect or a nearby child. It was followed by Aphex Twin techno beats and electronic squelches, as if R2D2 was on vocals. Was that one bird or two? I had no idea. Far more recognisable for me were the sounds of a bleating sheep in the distance, hissing car tyres on a busy road and a chugging tractor. Reverberations of the world from which I had come, and that much of my previous work had epitomised – stories and songs about motorways and flyovers and hospitals, eulogising the infrastructural. In this wild fen I felt like a stranger, like one of Hawkwind's intergalactic time-travellers, fleeing from his poisoned world to seek sanctuary on an alien planet that might be his own, but in a different aeon. I clutched my navigational smartphone with an outstretched arm, boots clacking on the ramparts that kept me aloof from the teeming swamp.

I might well criticise the venture capitalists who drained the fens for their own gain, but what was my role in this? After all, I was an outsider, draining the landscape of its stories, utilising its flora, fauna, topography and culture for my book, siphoning its raw materials to fill these pages for personal, artistic and financial purposes. When I was done, I would depart in a cloud of petroleum fumes, leaving the people of the Fens to continue their struggle against ecological collapse. The swoosh of rushes in the breeze swelled in volume as I passed, a whispering campaign against me and all my colonial bullshit.

The drive north-west from Wicken Fen took me through fields of rape and sunflowers, where pylon chain-gangs marched towards a scorched horizon and scarecrows were impaled on wooden stakes with their heads hung down. A billboard showed the beaming face of a local politician, asking for votes so that he could get the A10 to Ely turned into a dual carriageway. I wondered how much more this land could take, or what Hereward the Wake would make of the transformation since he had hidden out in the marshes where those low fields were now. The dual carriageway might seem to him the horrifying ultimate incarnation of those wooden ramparts laid by the Normans in their attempt to reach Ely and flush out the rebels.

A roundabout beside an Aldi car park slung me onto the Forty Foot Bank, an 8-kilometre-long raised road running along-side an artificial water channel. There was no wildlife on it, but for a solitary duck, looking ridiculous, is if it had mistaken the simulacrum for something real. A queasy horror gripped me. The road's straightness was unnerving. It kept going and going, through a barely changing landscape, trackways plunging to farm entrances on my right. Yellow speed cameras were perched on metal gibbets. Crows brooded on telegraph wires. Kestrels dotted in the sky barely moved, as if attached to a hanging mobile in a baby's bedroom. The outline of an enormous raptor came slowly into view but it turned out to be an artificial one on a wire, there to deter birds from the crops. As the wind battered my car on the exposed ridge, my knuckles whitened on the steering wheel. Numerous road signs warned of the dangers of plunging into the adjacent channel. One small spasm of my hand and I'd career down the bank to my death in the empty waters.

At the end of eight uneasy kilometres, the road turned sharply over a bridge, past the hulk of a Second World War pill box, hunkered on the bank, its hollow concrete eyes looking out for enemy boats from an invasion that never happened. I detoured down a country road from which I spotted an array of real raptors circling ploughed fields. One was so close that I was compelled to pull over. It was a marsh harrier, swooping over the sails of a ramshackle windpump at the back of a farm shed. I could see its fluttering feathers in intricate detail, caught in blazes of sunshine, then cast in shadow, then sunshine again, rippling from black to gold to silver. It looked fiery, like a phoenix. A ferocious elementary force above the spinning sails, unheeding of me, focused on whatever rodent scuttled below. I was listening to Hawkwind while watching a hawk in the wind. The synchronicity made me laugh. I wondered what a farmer might think if he saw a forty-seven-year-old man in a hatchback, cackling at a marsh harrier? That I was on drugs, most likely.

A light rain began to haze the fields as I entered Ramsey, a market town on a promontory projecting into the Fens. I pulled up on the high street outside the Windmill Bakery in search of a snack. The central reservation was dominated by streetlamps. Tall metal poles with two limbs outstretched. They mimicked the splayed arms of the wind turbines, which were just visible beyond the Victorian terraces: energy harnessers and energy emitters in a morphological duet. On the external wall of the bakery, a mural depicted an iconic fenland windmill beside a sunflower field with a butterfly, bee and ladybird. In the foreground was a crow. But most striking was what the artist had painted in the

background: a tidal wave at breaking point, towering over the scene. Of course, floods were once part of the natural cycle in the Fens. The nineteenth-century nature poet John Clare describes a winter flood as a series of furious waves that rise high like monsters who 'at the top curl up a shaggy main . . . then plunging headlong down and down'. But this mega-wave seemed to represent something singular and significant. Perhaps a past event, a local legend, or an intimation of a future disaster.

I had recently got my hands on a book by Christopher Marlowe, entitled *Legends of the Fenland People* (1926). Its musty sepia pages were filled with stories of dragons, saints and giants from oral lore, embellished – and in some cases possibly invented

– by the author.* In his foreword, Marlowe makes a bold claim that in the time of the ancient oak forest there was a 'mighty catastrophe in the shape of a seaquake, which caused immense tidal waves to sweep over the whole country and transform the area into a vast lake', which sounds more like one of the myths in his book than a historical fact. He writes that this cataclysm happened during the consulship of Valentinian, emperor from 364 to 375 CE. While there is no record of a tsunami, there was a dramatic change in sea level at that time. According to the writer Jim Hargan, the North Sea 'had a nasty little jump between 350 and 550 CE, flooding the coasts of northern Europe with an extra 2 feet of water and sending its inhabitants – folk known as Angles and Saxons – fleeing . . . into ill-prepared Roman territories'.[9] Undoubtedly this would have affected the regions of the Fens around the Wash and might have inspired the tale in Marlowe's book about Mandru and the Celtish rebellion against the Romans, in which their angry sea god sent a vengeful wave crashing down on the invaders, transforming the woodland into a wetland that the Romans could no longer control.

It is more likely that Marlowe's notion of a seaquake had its roots in the Storegga Slides, that mega-tsunami in 6200 BCE caused by a submarine landslide at the edge of Norway's continental shelf, which started Britain's island era. It would have sent a mighty wave surging across Doggerland and 40 kilometres inland through the forest at the north of East Anglia. As James Boyce points out in his book *Imperial Mud*, 'all the tree trunks preserved

---

* Marlowe insists that the majority of the contents is 'traditional'. The book contains a list of written sources for these legends and myths, including Felix of Dunwich, a monk who wrote about Saint Guthlac in the seventh century; William Hazlitt, the early-nineteenth-century essayist; and J. M. Heathcote, who wrote in 1876. Other than that, there were no references given to the individual legends, or an indication of how they might have been passed down.

in the peat [. . .] do face the same way', suggesting a catastrophic event. For a further clue to an ancient cataclysm witnessed by humans, he points to the local myth of Hrothgar, a fenland giant who did battle with the spirits of tempest and fire. To prepare for the big fight, Hrothgar created a gigantic dyke, as per instructions he received in a dream. As the wind blew down all the trees and the inferno raged, he let the water loose and it rushed down the channel through the Fens, extinguishing his enemies.

I noticed that on the bakery mural, just above the foaming crest of the tidal wave, the artist had painted a rainbow. It brought to my mind the rainbow in the story of Noah's flood, which represented an agreement between God and every living creature saved in the Ark that 'never again will all life be destroyed by the waters of a flood; never again will there be a flood to destroy the earth'. But in the case of the Storegga Slides, this promise might not be worth the parchment the Old Testament was written on. Professor Vince Gaffney from Bradford University's School of Archaeological and Forensic Sciences has studied the prehistoric event and believes conditions are ripe for a repeat. He says, 'The events leading up to the Storegga tsunami have many similarities to those of today. Climate is changing and this impacts on many aspects of society, especially in coastal locations.'[10] His colleague Dr Simon Fitch explains, 'One of the things [Storegga] has is gas hydrates, and they can become unstable when temperature rises. That would effectively cause another landslide.' If this were to occur, that vision of the tidal wave over the sunflower fields of the Fens might one day become a reality.

Whether there is another North Sea tsunami or not, the sea is rising and the Fens have never been more susceptible to flood. Much of the land I was driving through was below sea level. Only 8 kilometres north-east from Ramsey was the lowest place in the Fens – and, indeed, the whole of Britain. Holme Fen was originally a kilometre away from the edge of Whittlesey Mere, the shallow lake that once harboured so much plant and animal life, feeding and clothing the locals for centuries. The landowners behind the lake-drainage scheme were rightly concerned about the possibility of sinkage, which would make their new farms especially vulnerable to future flooding. In 1848, before the lake was drained using powerful new pumps, an engineer named John Lawrence piledrove a timber post through 6.7 metres of peat, embedding it in the clay beneath. This simple measuring tool would be their way of assessing the rate of decline as the project unfolded. As soon as water was pumped from Whittlesey Mere, the peat dried out and began to recede. Very soon, the wooden post poked through the surface. Month after month, more of it became exposed. A few years later, it was replaced with a hardier cast-iron version. Then a second post was added close to it in 1957. The land sank so much that they both became wobbly and required stabilisation with steel guys. Today they tower 4 metres high, marking a spot that is now 2.7 metres below sea level.

After the lake was drained, Holme Fen remained too wet and unwieldy, so it was left alone. Quickly, it became covered in silver birch trees. Today it retains areas of raised bog and flooded pits where peat has been extracted, teeming with life.* It is a rare example of resurrected wetlands, where the seasonal cycles are encouraged to flood and nourish the land once again. Unlike Wicken Fen, it was free to access. To enter, all I had to do was park on a country road near a train crossing and push through a wooden gate. I walked on spongey ground through the woodland, tangled with undergrowth and run through with water

---

* The Great Fen project aims to protect Holme Fen by connecting it to Woodwalton Fen, creating a 3,700-hectare expanse of wetland, restoring wild habitats and encouraging a return of the flora and fauna that have been lost.

channels. Some trees were felled and rotting; others splintered as if struck by lightning.

In *Waterland*, Graham Swift's acclaimed novel set in the Fens, he describes a soldier returning from the Great War, only to find that his homeland reminds him of the waterlogged trenches and bombed-out battlefields in Flanders. I could easily understand this, looking at the bark-stripped birches, blackened tree stumps and clumps of erupted earth. Exposed roots, coated in peat, were charred bodies in twisted death poses. Instead of clamouring artillery shells, a passenger train engine roared along a raised siding beyond the silver birches. The woods exploded with mechanical sound as the locomotive clattered through the scrub like a beast on the hunt for prey. Even when the noise subsided, it never quite went away. Within minutes, another train hurtled by, then another five minutes after that; a constant rise and fall, like the asthmatic lungs of infrastructure, wheezing in and out. But I could hear something else too – a pulsing drone – perhaps a distant buzz saw distorted in the breeze, a light aircraft circling, or some elemental distress signal from within the drained earth.

Deeper into the wood, the unrelenting monotony of silver birches became a blur, inuring me to what was really there. Rather than see the *woodland* I needed to see the *tree*. 'Silver birch' was merely the name of a species of vascular plant. But each one I passed by was uniquely scarred and flecked, in its own state of growth or decay, engaged in interactions with algae, fungi and insects. An ecological exercise I have learned is to consider an object like a tree not as an 'it' but as a 'thou': an individual being that has its own interrelations with other beings, its own destiny, its own form of consciousness even. This second-person perspective helps prevent us from seeing fens, woods and moors purely as resources with a utility for humans, and more as networks of inscrutable beings, interconnected in a delicate, sentient system that can never be wholly comprehended, controlled or broken down into parts.

'Hello, you,' I said to one of the trees. She was gleaming white, tilted back as if punch-drunk, a single root poking out of the mud like a hernia and a knot halfway up her trunk in the shape of an eye. At this point I realised I had not spoken to another soul that day. As far as I could tell, I was alone. But it didn't feel like it. I grew paranoid, alert for sudden surprises that might disrupt my isolation or suggest that unseen entities watched me from the undergrowth. As I passed waterlogged ditches, their glassy surfaces trembled, as if a tyrannosaur was approaching. Rationally, I knew this was the movement of aquatic insects but the effect on my peripheral vision was like the onset of an epileptic fit.

As I proceeded to the far side of the wood, passing camouflaged bird hides on spindly struts, the buzzing drone got louder, now undulating like a foghorn. I came upon an expanse of reedy marsh. It seemed as if the sound was moving back and forth along the horizon, but I couldn't see anything in the sky. How did I know that I was not having a UFO experience? Or that I wasn't receiving test signals from a diabolical occult experiment? Or that this wasn't all a projection of my troubled mind? Suddenly, with the kind of timing a horror director would kill for, a Chinese water deer crashed from the reeds right before me. I cried out, heart in my mouth. These stumpy-legged beasts with vampiric tusks had been introduced to Britain in the nineteenth century and kept in captivity until escapees from Whipsnade Zoo took their species into the wild. They were well established in this borderland between the shattered birch boneyard and the reinstated marshlands. But this one startled me, all the same.

'Goodbye, you,' I said, under quickened breath, as she vanished into the sunken land.

I journeyed north through an area of Lincolnshire known as South Holland, the 'heart of the Fens', which borders the Wash. The flatland never seemed flatter. Ditch networks formed angular grids of fields, ploughed into trapezoid ridges, like muddy slabs of corrugated iron, bordered by fences and hedgerows. There were so many straight lines. Everywhere I looked, I was confronted by what Graham Swift described as the Fens' 'intolerable geometry'.[11] In one field, scarecrows dressed in white spacemen suits looked like visitors from the future, returned to the Earth of their deep past, seeking answers about the cause of its destruction in this drained plateau, watching me pass in my car, remembering legends of petroleum machines and how a great plundering of the land was required to keep their engines turning.

Eventually, all the flatness terminated at a long green embankment, which formed a barrier between the land and the sea. This was Gedney Drove End. It seemed like the rim of the world, a ledge beyond which nothing might exist. The Romans once built embankments here, like this one, to protect agricultural land from flood, only for them to crumble, along with their empire. During the Second World War, the Wash's sea banks were repurposed as defensive lines, dotted with pill boxes. Now they were in service again for the protection of the country against elementary forces: an inexorably rising sea at the edge of a shrinking land.

On top of the ridge, two human figures shimmered in a heat haze. I watched them while I stopped my car by a hawthorn thicket on the edge of a stony field. I was perturbed. The figures didn't move. Perhaps they were watching me. Perhaps I was not meant to be here. I had just passed some signs for an MOD firing range. Now I was worried that I had parked in a forbidden zone and they were staring at me in horror, waiting for my head to explode with shrapnel from a rogue round. But as I walked closer, I realised that they were not people at all, but fence posts. The Fens were warping my perception, giving me visions of a world where the animate and inanimate had become indistinguishable. But perhaps there was no real difference. At the smallest scale, everything in existence is made of atoms that are 99.99 per cent empty space, with electron particles that appear and disappear in different positions around a nucleus in a form of inexplicable quantum magic. That goes for rocks, tin cans and fence posts as much as for butterflies, bees and people. When exactly can something be said to be alive? What is the precise point when seemingly dead matter suddenly becomes conscious life? It has been argued that the universe, and everything in it, is conscious, or has the potential for consciousness, and that we humans are not separate entities but part of a single, entangled quantum whole.

I ascended the steep bank and looked over a saltmarsh, thick with mud, cracked with water channels. Stagnant pools reflected the blue sky between patches of samphire, pungent with salt and decay. Beyond the marsh was the sea, where the great oak forest once would have been. A solitary fishing trawler drifted between buoys on the calm water. The air resounded with cries as geese flapped across the skyline towards the coastline of Norfolk, a dark undulation on the eastern flank of the bay.

On a beach at the foot of that distant headland, a ring of Bronze Age timbers was exposed after storms in 1998. Its fifty-five oak posts surrounded an upturned tree stump with roots splayed to form an altar-like centrepiece. After the land became waterlogged, the structure was subsumed into peat, then covered in sand as the sea advanced. Some believe that 'Seahenge' was used for sky burials, where a corpse was left on the altar to be eaten by birds. It was excavated and carted off to Flag Fen by English Heritage, angering Druids, who believed it should remain in situ. Before its departure, multimedia artist James P. Graham shot timelapse footage on a Super-8 camera. His three-minute film, *Losing Seahenge*, begins at low tide, with the timber posts fully exposed. The tide comes in, consuming the altar, as the crackling film reel decays into whiteout, obliterating not only the monument but the observer, implying the threat of our own disappearance along with that of Seahenge. The artist described it as an 'allusion to the potential advent of ecological disaster and annihilation'.[12] The floods that took the Bronze Age site might eventually take us, too.

Directly ahead of me in the Wash was a strange island – a disc with a flat top and uniform slopes to each side. There was no vegetation. No structures. I checked the map on my phone to work out what it was. But nothing was marked on there. I felt the same discombobulation I experienced on the balcony in Llandudno, North Wales, staring at the phantom illuminations of a coastline that shouldn't have been there. Perhaps the island was in my imagination, a mirage, or yet another mythopoeic signal from the deep past. I would later discover that this was the Outer Trial Bank, constructed in the 1970s to test the feasibility of a freshwater reservoir by trapping the outfall of the rivers that emptied into the bay. The scheme was abandoned but the island has become a valuable nesting ground for birds. As I contemplated the weird isle, a swarm of large black flies with dangly legs thwacked against my jacket and followed me in an unruly insect escort as I tried to escape. These were St Mark's flies, which hatched on St Mark's Day, 25 April. I was struck by the synchronicity: St Mark was the patron saint of Venice, built on islands in a marshy lagoon at the edge of the Adriatic Sea in north-east Italy. The legend is that when St Mark arrived at the lagoon, an angel appeared and told him, 'Peace be with you, Mark, my evangelist. Here your body will rest.' Like the Fens, its settlers found refuge in the wetland, where there was plenty of wildfowl and fish. The city was founded in the fifth century, the same time as the events in the Lowland Hundred and Tyno Helig myths. Similarly, the first Venetians drained the land, digging canals and building embankments. Today they face a similar fate. The weight of the city is sinking it into the mire and estimates are that Venice will be under water by the year 2100.

The threat to the Fens couldn't be clearer from my vantage point on the embankment. I could see the saltmarsh and sea on one side, and fields on the other, much lower down. This elevation was the only barrier between the sinking land and the rising ocean. All those quaint hamlets, vicarages and country pubs; all those fields of tulips, potatoes and corn; all those windmills and pylons – they were predicted to be underwater by 2050, should current rates of global emissions continue unabated. In that year, in this same spot, I might look south across an inland sea stretching all the way to Ely, once more an island, and Ramsey, perched on its promontory above the waves. It would look exactly like the aftermath of the flood in the story of Mandru and the Romans, or the tidal wave described by Christopher Marlowe in his introduction to *Legends of the Fenland People*.

But would it be such a disaster in environmental terms? After all, this land is supposed to flood. Almost 200 years ago, John Clare described the Fens in wintertime like this: 'From one wet week so great an ocean flows / That every village to an island grows.'[13] These flood events weren't catastrophic for the fen dwellers who had learned to live with natural seasonal cycles and sustain themselves from the bounty that resulted. It was only because of our incessant drive to control, contain and exploit wetlands like these for profit that I could park my car on what now should be seabed and bear witness to this doomed agricultural landscape.

The story of the Fens shows that we can never hold back the waters permanently; the more we dig in and refuse to adapt to the natural flux, the more acute the disaster we shore up for ourselves. Perhaps one day this sunken land will rise again, like Seahenge, exposed to the light after millennia of submergence beneath the sand. But what form it might take in its new incarnation, and whether we might be around to walk on its resurrected bones, is another question entirely.

5

ATLANTEAN DREAMS

At the end of the Ice Age, Atlantis was a powerful island state, with great influence over the world, exporting luxurious goods and cultural knowledge. Its capital city was situated by the sea on the edge of a fertile plain, surrounded by mountains. Hot water ran from thermal springs through a sophisticated plumbing system. Boats loaded with produce travelled through a network of canals to and from the busy harbour. Balloons made from elephant skins, filled with natural gas, allowed Atlanteans to fly around the world and journey beneath the sea. Giant domes channelled the sun's energy as a power source to illuminate the interiors of multi-storey buildings. A magnificent temple boasted columns of onyx and topaz, inlaid with beryl, amethyst, and stones that caught the sun's rays. Inside, they sacrificed bulls to appease the gods and ensure that their progress could continue.

Atlantis's abundant natural resources and technological prowess helped its society remain harmonious and virtuous. For centuries it was a place of high civilisation, divided into social classes and trades, with some dwelling in the cities and others in idyllic farmland. However, their constant harnessing of gas and radiation created geological instability. A series of volcanic eruptions transformed Atlantis from a single landmass into an archipelago, with Poseidia, Aryan and Og as its largest islands.

The Atlanteans tried to adjust to their diminished circumstances, but all was not well. Their powerful elites became increasingly arrogant and greedy, warring with other nations and plundering their natural resources, while the populace grew indolent and quarrelsome.

This decline in moral virtues greatly angered the gods, who destroyed Atlantis with a violent earthquake. As the walls fell and the temples collapsed, the waters rushed in and the islands sank in a single day to lie at the bottom of the ocean, where their ruins were covered over by sand and eventually forgotten.[1]

Hampton Roads, Southeast Virginia, 3 February 1932. A balding man in his late fifties lay on a couch in his office on Virginia Beach. His tie was loosened around his neck. Shoes off. Eyes closed. Hands clasped over his stomach. He breathed deeply until he appeared to fall asleep. Around him sat onlookers, including his wife Gertrude; members of the Association for Research and Enlightenment; and stenographer Gladys Davis, fingers poised above the keys, ready to transcribe his words. After a prompt from Gertrude, the sleeping man spoke in a low monotone about the continent of Atlantis. He acknowledged that the legend was based on a story told by Plato, but scientists believed 'such a continent was not only a reasonable and plausible matter, but from evidences being gradually gathered was a very probable condition'. This was the first in a series of psychic explorations in which the sleeping man revealed details of Atlantis's technologies and architecture, its ruination and submergence, and the sunken hall of records that would rise again for the enlightenment of humankind.

Edgar Cayce grew up in rural Kentucky. As a boy he regularly spoke to his dead grandfather, played with invisible friends, and saw auras around people. Legend has it that he would fall asleep with his head on his schoolbooks and awaken with a photographic memory of their contents. When he was a young adult, he suffered paralysis of the vocal cords. Desperate to find a solution, he put himself into a trance and requested that a doctor ask his subconscious questions about the problem. The remedies prescribed as a result cured him within a year. Afterwards, he continued to put himself into dream states, where his mind was free to roam into what he described as a 'universal consciousness', unshackled from time and space. In this fourth-dimensional plain, the thoughts and deeds of individuals – living, dead and yet-to-be-born – were imprinted like audio recordings. Through somnolent meditation, Cayce believed he could access this information and play it back.

Until his death in 1945, he gave thousands of psychic readings to knowledge seekers who asked him questions about a wide range of subjects, from health, parenting and spiritual matters to past lives, ancient history and supernatural phenomena. In response, he would channel the universal superconscious into meandering monologues, faithfully recorded by a stenographer. He became known as the 'sleeping prophet' and 'miracle man of Virginia Beach' for the accuracy of his visions.

One of his prophecies was that in the twenty-first century the planet would endure cataclysmic events that would radically reconfigure society. These 'Earth changes' included earthquakes, melting ice caps, floods, solar flares and a reversal of the magnetic poles. Major cities on America's East Coast would disappear beneath the sea, along with the southern portions of Carolina and

Edgar Cayce on the couch used for his readings.

Georgia, and everything in the United States west of Nebraska. Despite these dire predictions, Edgar Cayce had been instructed by his own subconscious, during a trance, to move to Virginia Beach in 1925. This was one of the spots on the East Coast most vulnerable to the geological shifts and sea-level rises of which he warned, for the land was sinking, thanks to events that began 20,000 years earlier.

During the Last Glacial Maximum, the Laurentide Ice Sheet covered Canada and the northern parts of the USA, reaching as far down as Illinois and Pennsylvania. The glacier pushed down on the land and forced up the mid-Atlantic coast, south of its terminus, like the opposite end of a seesaw, creating a bulge. After the ice retreated, isostatic adjustment caused the eastern Atlantic to sink back down. One of the most affected places is Hampton Roads in the Tidewater region of south-east Virginia, an estuary of tidal marshes and swamps, where rivers feed into the Chesapeake Bay and the Atlantic Ocean. Not only is it sinking because of post-glacial geological rebound, but also because of the amount of water pumped from the ground for use in households, businesses and agriculture. Levels have dropped 60 metres in a hundred years. Like the English Fens, this land is drying out and shrinking. Meanwhile, the sea level is rising by 2.5 centimetres every four years. 'Where you have two millimetres per year of sea-level rise, another one to three millimetres per year from compaction is pretty substantial,' says Jason Pope, a hydrologist from the US Geological Survey.

The state has plans for stormwater and sewage-system protection, along with a scheme for injecting water into the ground. But scientists predict another 15-centimetre rise in sea level by 2035. Should a hurricane strike during a tidal surge, swathes of the shoreline would flood. There is an evacuation plan for residents in Virginia Beach and the city of Norfolk to escape through the Hampton Roads Bridge-Tunnel, evoking those mythic causeways built by King Gwyddno for the similarly beleaguered citizens of the Lowland Hundred. And we know what happened to them.

When asked if Virginia Beach would sink, Cayce was surprisingly positive. 'Of all the resorts that are in the East Coast, Virginia Beach will be the first and the longest-lasting,' he declared. 'The future is good.' This might not go down as one of his most accurate prophecies. Unless the universal superconscious is privy to information we don't yet have, the situation doesn't look good at all. Based on current rates of subsidence and sea-level rise, Hampton Roads might face the same fate as Atlantis, one of Cayce's primary subjects.

Throughout 1932 and 1933 he devoted a series of readings to the lost continent for the benefit of a research group. The transcripts relate an Edenic origin story in which the first Atlanteans were more like thought forms – expressions of the universal consciousness – than individuals, with no difference between male and female. As they manifested physically, they split into genders, developing individual personalities. An increase in ego led to a separation from the divine and a desire for technological advancement through occult science. When Atlantis was destroyed, their houses, temples and statues sank to the bottom of the ocean. However, Cayce claimed that there was a yet undiscovered hall of records in Egypt, housing Atlantean secrets. Furthermore, three temples would re-emerge, including one near Bimini, a chain of islands in the Bahamas. Thirty-five years later, a diving team led by archaeologist Joseph Mason Valentine discovered what appeared to be a paved road of huge, flat limestone blocks, 6 metres deep, near North Bimini Island. There was speculation that this was the work of an ancient civ-ilisation. However, the scientific consensus is that the 'Bimini Road' is a natural feature created by tidal erosion, which scours lines into the limestone. But even if this is the case, it remains highly strange that the prophet of Virginia Beach chose that very location as the site of an Atlantean temple.

At any mention of Atlantis being a real place, archaeologists are likely to raise an eyebrow. The existence of a civilisation in the Ice Age would entirely upend the history of humanity as we know it. The most widely accepted theory is that civilisa-tion began after the Ice Age, when climate instability forced the Mesopotamians to develop agriculture so that they could stabilise food supplies and trade the surplus, leading to innovations in writing and architecture, along with the foundation of city states and organised religion.

However, there have been intriguing discoveries that challenge this narrative. The temple complex of Gobekli Tepi in Turkey, for instance, built at least 12,000 years ago, in which symbolic animal carvings with possibly astrological significance cover enormous T-shaped limestone pillars. It's a site that long predates the

Neolithic, when organised rituals in permanent stone structures were presumed to have begun. There are other anomalies, including the Pantelleria Vecchia Bank Megalith, a 15-tonne limestone block 12 metres down on the seabed between Sicily and Tunisia, drilled with holes of consistent diameter. Based on estimates of sea-level change, it would have last been above the water 10,000 years ago. Off the coast of Cuba, what appear to be pyramids loom on an area of seabed that would have been dry land 50,000 years ago. Most intriguingly, in 2015 scientists discovered native Australian and Melanesian DNA in some indigenous people living deep in the Brazilian Amazon, and yet none of those same genetic fingerprints is present in North America. This has led to conjecture that there might have been an inter-continental oceanic crossing as early as 15,000 years ago, which not only contradicts the established story of human migration but suggests greater knowledge, skills and societal organisation than is believed possible at that time.

Most archaeologists argue that there cannot have been civilisations in the last Ice Age because there would be more evidence of their structures. Defendants of the lost-civilisation theory counter-argue that anything that existed before the Younger Dryas climate disaster – whether it was caused by a comet, a mega-flood or sudden disruption to the Gulf Stream – would have been destroyed by glaciation, earthquake, fire and floods. If between 60 and 80 per cent of large mammals died out in Younger Dryas, then the same ratio might have applied to humans, which means most people on the planet perished, along with their cultures and societies. Settlements would have been predominantly coastal, with sea levels hundreds of metres lower, so the majority would be deep under water and sand by now.

Putting archaeological controversies to one side, the dark side of ideas about lost advanced civilisations such as Atlantis is that they can play into the hands of racists who reframe them as being the domain of a superior early European or white people. It encourages the notion that indigenous cultures cannot have been responsible for their innovations in architecture, mathematics and astronomy, which must have been delivered by super-advanced races. For instance, the Nazis were so interested in the idea of Aryan Atlanteans that Heinrich Himmler formed a Bureau of Ancestral Heritage in 1935 to seek out the descendants of the flooded city. They explored the Himalayas, where they believed the Atlantean survivors had fled after the flood, and began measuring the faces of Tibetans in an attempt to ascertain their racial origins.

However, what is of interest to me is the way that the Atlantis legend illustrates how geological and climatic events influenced our religions, philosophies, art and politics, and how they might help us understand what is happening on our planet today. Whether it existed or not, Atlantis is a living cultural artefact, exerting its influence through cinema, literature, music, journalism and social media. To explore its imaginal landscape is to travel a curiously shifting geography of the collective consciousness. The myth has mutated through epochs, taking on different meanings and expressing new anxieties, while retaining a universally recognised core narrative. It is almost always referenced in articles about discoveries of sunken archaeological sites and, after Noah's Ark, is the most pervasive flood myth. Atlantis has featured in numerous books, including Arthur Conan Doyle's *The Maracot Deep*, Ayn Rand's *Atlas Shrugged* and Robert Shea and Robert Anton Wilson's discordian classic *The Illuminatus Trilogy*. Millions of children have encountered it through the *SpongeBob SquarePants* episode 'Atlantis SquarePantis', the animatic Disney movie *Atlantis: The Lost Empire*, the superhero film *Aquaman* and the video game *Tomb Raider*. Most people would be able to relate the bare bones of the story – an ancient civilisation that sank beneath the waves – even if they know nothing of its origin.

It was in 360 BCE that the Greek philosopher Plato travelled to Egypt and afterward wrote a semi-fictional dialogue known as *Timaeus*. It told of how, 300 years earlier, the Athenian statesman Solon visited the town of Sais in the western Nile delta, a last vestige of the ancient Egyptian civilisation. In a temple, he spoke to a priest about the history of Greece. The priest mocked Solon's poor comprehension of the past. He told him that Athens was older than he could conceive, with roots in an unremembered time before known history. There had been 'many destructions of mankind arising out of many causes,' explained the priest, 'the greatest having been brought about by earth-fire and inundation.' After each annihilation, humans began again, like children, without knowledge of what happened before. Written accounts of the most recent inundation had perished but there were secret records, concealed in their temple, of an ancient Athens that existed 9,000 years before Solon's birth. It was at war with a powerful island nation, known as Atlantis. He described this civilisation in detail to Solon, explaining that the island was eventually sunk by wrathful gods, who destroyed it over the course of a single day through earthquake and flood.

While Plato's dialogues used fictionalised elements, the philosopher insisted that this account of Atlantis was fact, not legend. His description dates it at around 9600 BCE, the period of runaway global warming, floods and geological stresses at the end of the Ice Age. These events were experienced by societies advanced enough to build ships, trade goods overseas and construct temples like the ones discovered at Gobekli Tepi. It is therefore highly feasible that important trading posts, settlements and sites of worship were destroyed, along with the knowledge and culture of those communities. Their scattered populations might well have taken their knowledge to new regions. For instance, ancient Mesopotamian cultures revered an amphibious figure named Oannes who came ashore in the Persian Gulf with a cohort of sages known as the Apkallu. They spent their days teaching people writing, mathematics, law, farming and construction: a blueprint for civilisation from a lost world. This story survived thanks to the writings of the Babylonian priest Berossus in the third century BCE. He claimed that his sources were from the temple archives of Babylon, which contained public records going back 150,000 years. We cannot know for sure if Oannes was based on real émigrés from an Ice Age flood disaster but there are other intriguing connections between mythic literature and geological and climatic events.

India suffered a huge loss of land after the Last Glacial Maximum, particularly on the west coast and in the Bay of Bengal. The *Mahabharata*, an epic written around 300 BCE, describes events that purportedly happened a thousand years earlier, although its source material might have come from oral traditions reaching further back in time. It is the epic story of a power struggle between two families, the Kaurava and the Pandava, and features the lost city of Dwarka on the coast of the Saurashtra peninsula in north-west India. This was likely to have been an important hub in the oceanic trading nexus – and its name is translated as 'gateway'. Dwarka was said to be the home of Krishna, an earthly incarnation of Vishnu who was forewarned of a flood disaster and persuaded the city's inhabitants to leave before the deluge. The *Mahabharata* describes how the people departed in a calm procession as the floodwaters swelled behind them. But they didn't lament their cursed fortune. Instead, they accepted the necessity of the flood and said, 'Wonderful is the course of fate.' They believed that golden ages and dark ages revolved in cycles, and everything must eventually return to the beginning.[2]

How much of the *Mahabharata* is based on true events, and when they might have occurred, is disputed. But underwater archaeological investigations of Dwarka's supposed location have uncovered the foundations of pillars, irrigation systems, jetties and a port wall. There have been further findings of a submerged city in the nearby Gulf of Cambay, with human bones, pottery and a piece of wood carbon dated to 9,500 years ago, predating the oldest known Indian city by 5,000 years.* This would place it in the time of the post-Ice Age floods after 9500 BCE. Its people might have witnessed extraordinary climate disasters and told stories about it, which were handed down the ages, mutating into new forms, reflecting the hopes and fears of the cultures that inherited the narrative.

In the medieval period, Atlantis became of interest to geographers who suspected that there were bounteous, undiscovered lands across the Atlantic. After Columbus stumbled on the Bahamas, the Americas were considered Atlantis by some scholars, its Mayan and Aztec ruins suggesting a forgotten ancient civilisation. For white European colonisers, the Americas were a blank slate, granted to them by God, devoid of any purpose but for their utility, to remake in their own image. The New World offered an opportunity to experiment with new models of society, both in reality and in fiction. In the sixteenth century, Thomas More envisioned the lands of the New World as a utopia, and in 1626, Sir Francis Bacon's *New Atlantis* framed a mythical island near Peru as a place of advanced science and high morality. Other writers used the mythos of Atlantis to suggest that racially superior white ancestors had bestowed on the peoples of the New

---

* Dilip Chakrabarti, a Cambridge University historian, says, 'If the dates are true, it would be revolutionary in terms of understanding the growth of villages and cities in the world.'

World their architecture and culture. These Renaissance versions of the legend were fundamentally an expression of colonial prejudices and anxieties.

The legend took on renewed vigour in the industrial age, when great cities bristled with smoking chimneys, locomotives tore through the hills and passenger ships traversed the seas. As life became increasingly mechanised, there were fears about where our society was heading and what might happen if we lost control. Accordingly, the late nineteenth century saw an increased yearning for spiritual knowledge and the exploration of psychic phenomena, testing the boundaries between the known and unknown, seen and unseen. In her writings, Madame Helena Blavatsky, founder of an occult organisation, the Theosophical Society, disputed the myth of progress: that history moved like an arrow from ignorant savagery towards technological utopia. She subscribed to the idea that history was cyclical. Civilisations rose and fell. Her belief was that all humans possessed within them the secret knowledge of the universe, but that it was hidden from our egos and rational thought processes. Only through conscious effort and spiritual practices could we remember that which had been forgotten. In her book *The Secret Doctrine*, she related an alternative history in which 'root-races' had existed in epochs lost to memory. Each incarnation of humanity represented a stage in our spiritual and physical development. The fourth 'root race' existed on Atlantis in a golden age of knowledge. But the Atlanteans' moral decline and abuse of technology brought about their destruction through earthquake and flood. The survivors sailed across the world to plant the seeds of their knowledge in new civilisations. So began another revolution of birth and death, in which humans could acquire greater spiritual powers, and so it would continue, cycle after cycle, until we realised our divinity.

Like the trance-induced transmissions received by Edgar Cayce, Blavatsky's book was supposedly telepathically dictated to her by adepts she had encountered on her travels in Tibet – curiously, the same place Himmler's Nazi scientists would go on to search for Atlantean descendants. But occult accounts such as Blavatsky's weren't necessarily literal histories. As the occult scholar John Michael Greer points out, her Atlantis was an example of what the mystic and philosopher G. I. Gurdjieff called a 'legominism': a method of encoding spiritual ideas in stories that could survive long periods of time, through extreme societal changes.[3] On the surface, they could be enjoyed by anyone but, for occult practitioners, deeper truths could be decoded from them.

With a similar aim, the polymathic occultist, magician, author and mountaineer Aleister Crowley wrote *Liber LI: The Lost Continent*. In his version of the legend, Atlantis was an archipelago. At its centre was a pillar of rock, residence of a god-like being named Atla. In this society, servants and lower classes were ruled over by hirsute magicians with psychic powers, who experimented with a substance known as Zro, extracted from the sweat of servile workers. This allowed them to transmute into new forms through orgiastic ritual magic. They practised 'dreaming true', unexplained by the narrator but suggesting that, like Edgar Cayce, they accessed gnostic truths through trance states. The ultimate goal of the Atlantean magicians was to transcend their earthly bounds and ascend to Venus, from where they would launch into the universe. Atlantis was destroyed after their experiments with Zro created a divine manifestation in the form of a human child. At this point, the column of stone containing Atla lifted off like a rocket and flew to Venus. The archipelago was abandoned to earthquakes, tidal waves and flood as its survivors fled to Egypt, where they were received as gods.

While Crowley's Atlantis was an intimation of humankind's potential to ascend to the stars, the pioneering horror author H. P. Lovecraft posited Atlantis as something dark and unknowable from long before us. In his story 'The Temple' (1920), a damaged U-boat sinks in the North Atlantic. The narrator, a German officer, sees the ruins of a city on the ocean floor, with temples and villas, marble columns and sculptures of 'terrible antiquity'. He realises that this is Atlantis, 'a culture in the full noon of glory when cave-dwellers roamed Europe and the Nile flowed unwatched to the sea'. Lovecraft took this idea further in *The Call of Cthulhu* (1928), in which a telepathic cosmic entity lurks in 'the nightmare corpse-city of R'lyeh' in the South Pacific. The city was built in 'measureless eons behind history by the vast, loathsome shapes that seeped down from the dark stars'. These beings had come from civilisations that existed billions of years ago. On Earth they built cities and fought battles, until R'lyeh was submerged with Cthulhu trapped within it. Lovecraft gave its coordinates as 47°9 south and 126°43 west, the point in the Pacific farthest from any landmass.

It was in this very spot that the National Oceanic and Atmospheric Administration's listening stations picked up a strange ultra-low frequency sound in 1997. People wondered what might have emitted this animal-like noise, known as the 'Bloop'. Perhaps Lovecraft had correctly intuited the presence of something monstrous in the depths. Years later, seismologists announced that the sound was from a breaking ice shelf in Antarctica. While this might have disappointed those seeking proof of subaquatic gods, it illustrates how we continue to be haunted by geology. Legends including that of Atlantis are, in part, attempts to explain mysterious topographies, created by awesome forces beyond our knowledge in chronological scales that are difficult to conceive.

In the nineteenth century, scientists began to realise that the world was a lot older than hitherto had been imagined and had not always looked the same. There were anomalies in the distribution of rocks and animals that couldn't easily be explained in 1864. British zoologist Philip Lutley Sclater wondered about the lemurs of Madagascar, and how there could be fossils of them in India, to the east, but not in Africa to the west. He surmised that there must have been a land bridge between the two landmasses, which allowed these animals to extend their territory. Sclater proposed that a lost continent extended 'across the Indian Ocean and the Indian Peninsula to the further side of the Bay of Bengal and over the great islands of the Indian Archipelago'. He called this Lemuria.

This theory inspired Ignatius Donnelly, a Minnesotan lawyer, to write *Atlantis: The Antediluvian World* in 1882. He placed Atlantis in the Atlantic, where he believed a landmass like Lemuria to have existed, with Egypt as one of its colonies. Its leaders' power struggles, trials and tribulations became the origin of ancient Greek and Norse gods. Similarly, in 1924, the Theosophist Lewis Spence postulated that Atlantis was in the eastern Atlantic, linked to Europe via a land bridge destroyed by earthquakes around 22,000 BCE. The remaining island was a powerful empire until it was shattered by further earthquakes in 9600 BCE. All that was left above water were its mountain tops, which we know today as the Canary Islands and the Azores.

After the discovery that Earth's crust was formed of shifting tectonic plates, which drifted apart, or collided, over millions of years, the theory of Lemuria was discredited. But it remained clear to scientists that large tracts of land around the world had sunk – and were still sinking – because of isostatic movements, volcanic activity and sea-level changes, particularly around the coast of India, Australia and north-west Europe. The discovery of Doggerland proved the existence of a land bridge to modern Denmark, Germany and the Netherlands. Another between Cornwall and the Isles of Scilly was suspected to be the location of the sunken kingdom of Lyonesse. In south-eastern Asia, a peninsular shelf known as Sundaland was inundated by seas 11,600 years ago, flooding its plains and river deltas. Its remaining highlands form the Indonesian archipelago. Geologist Danny Natawidjaja has spent years researching what appears to be a buried pyramid on the megalithic site of Gunung Padang on Java, which he thinks might be more than 12,000 years old. He believes therefore that Indonesia might be Atlantis, or at least some part of it that is now under water.

Another candidate for Atlantis is in the Aegean Sea, where the Minoan civilisation flourished on the island of Crete during the Bronze Age. The Minoans built cities with palaces, temples and tombs, connected by roads and serviced by plumbing systems. Elaborate frescos adorned their homes, depicting natural landscapes and wild animals. Their jewellery, pots and vases were traded across the Aegean and Mediterranean, to Egypt, Palestine and Turkey, spreading their influence far and wide. Just like the Atlanteans described by Plato, they worshipped bulls, and specialised in crafting objects from metal. On the island of Santorini, the location of Thera, an active volcano, they built the city of Akrotiri. Around 1600 BCE, Thera erupted, triggering a tsunami that pounded the surrounding islands with a wave 35 metres high and 15 kilometres long, devastating the coastline of Crete.* As the volcano spewed out rock, gas and ash, the magma chambers

---

* Research in 2008 showed evidence on the Crete coastline of 'chaotic layers of material including volcanic ash, building material, pottery shards, marine shells, bones and beach pebbles', which appeared to be the result of strong water currents. (Bruins et al. 2008, *J. Archaeol. Sci.* 35, pp. 191–212.)

beneath it emptied out and Santorini sank. All that remained were three small islands around a flooded caldera. This was one of the largest eruptions ever witnessed, and many believe that the event survived in the collective memory, entering Egyptian historical accounts, from which Plato picked it up for his Atlantis allegory.

After the eruption, the Minoan capital city of Knossos on Crete remained intact, but the weakened civilisation became vulnerable to trading competition and attacks from rivals. Within a few hundred years it collapsed as the warlike Mycenaeans established their dominance. By 1100 BCE the Mycenaeans had fallen too, along with other late Bronze Age civilisations, including the Hittites and the New Kingdom of Egypt. A team of archaeologists, geologists and historians researching submerged ancient beaches in the Mediterranean have found evidence that there was a period of climate change after Thera's eruption. A rapid cooling of the Mediterranean Sea reduced the region's rainfall, instigating a drought that continued for four centuries, leading to famine, disease and warfare, tipping established societies into chaos.

This effected a profound change in human consciousness, according to the psychologist Julian Jaynes. In 1976 he proposed a controversial theory that until around 3,000 years ago, the mind operated differently than it does today. The left side of the brain would respond obediently to the right side of the brain, which provided godlike commands in the form of an inner voice that people literally *heard*. He called this the 'bicameral mentality'. Humans at that time were not naturally introspective. They didn't have the capacity to reflect on or question their consciousness, which provided important information as if from an external source. However, this mentality changed after the collapse of late Bronze Age civilisations in the Mediterranean. 'The second millennium BC was heavy laden with profound and irreversible changes,' wrote Jaynes. 'Vast geological catastrophes occurred.

Civilizations perished. Half the world's population became refugees. And wars, previously sporadic, came with hastening and ferocious frequency as this important millennium hunched itself sickly into its dark and bloody close.'[4]

Jaynes argued that in waves of mass migrations, scattered populations from collapsed civilisations faced unpredictable new situations to which they had to improvise and adapt. To cope, we evolved a mindset that allowed us to assess complex scenarios, share experiences with strangers, and respond in a more flexible way than simply by obeying inner commands that no longer correlated with reality. Climate change killed the old deities and left us to our own devices. 'In social chaos the gods could not tell you what to do,' wrote Jaynes. 'Or if they did, they led to death, or at the intimate least to an increase in the stress that physiologically occasioned the voice in the first place.' If he is correct, this neurological adaptation became endemic, conditioning how we think today. It is intriguingly resonant with the idea of the Atlanteans' separation from the divine, expressed in many occult versions of the legend as the cause of their fall.

While Jaynes' theory is not universally accepted, psychiatrist and neuroscientist Iain McGilchrist agrees that a significant brain change did take place during ancient Greek times. However, in his book *The Master and His Emissary: The Divided Brain and the Making of the Western World*, he proposes that it involved an increased separation between the left hemisphere (which specialises in abstraction, language, philosophy and mechanics) and the right hemisphere (the hub of intuition, instinct, empathy and emotion). This has led us today to a dominance of left-brained thinking, which sees the world – and the human brain itself – as a machine that can be categorised and broken down into parts, rather than as a fluid, interconnected and mysterious holistic whole. We no longer heed the voices of our right hemisphere and believe only in what can be seen, measured and utilised for our technological advancement, leading us into a dangerous new epoch of anthropogenic climate change, just as the legendary Atlanteans' technology led them into trouble with the old gods.

These notions of an adaptive brain change shows us how geological and climatic shifts can birth new ideas, new forms of consciousness, new ways of living. I wonder what might emerge from our own period of environmental collapse, which will undoubtedly cause waves of migration and desperate resource wars as the world is reshaped.

The UN's environment agency has said that there is 'no credible pathway' to the goal set by the Paris Agreement to limit global warming to 1.5°C. Our failure to reduce emissions means that the only option left is a 'rapid transformation of societies'. It is estimated that there are enough greenhouse gasses in the atmosphere to lock in over 12 metres of sea-level rise. The Arctic is warming three times faster than anywhere on the planet and could be sea-ice free in the summer by the 2030s, a decade earlier than predicted. This will switch off Earth's cooling systems, with less white ice reflecting the sun and more dark water absorbing heat, accelerating warming and driving up sea levels. Melting Arctic ice is weakening the jet stream that conveys winds from west to east, leading to more extreme weather in North America, Europe and Asia. Meanwhile, the East Antarctic Ice Sheet, which contains 80 per cent of the world's ice, is also melting fast. In 2022, a chunk the size of Rome broke off after temperatures 40°C warmer than usual at that time of year. The disappearance of this ice will slow the deep 'southern ocean' current, altering the climate and increasing global rainfall.[5] Scientists are particularly concerned about the Thwaites glacier, known ominously as the 'doomsday glacier'. It is the size of Florida, and should it collapse, the seas will rise by half a metre almost instantly. In a theoretical scenario, if all the ice on Greenland and the Antarctic were to melt overnight, seas would rise by 66 metres, sinking large parts of Bangladesh and India, the Nile and Mississippi deltas, the Amazon basin, the Netherlands and much of northern Europe: Shanghai, New York, London and Bangkok – all gone. Almost half the world's population would need to flee to higher ground.

These are the kind of radical Earth changes that Edgar Cayce prophesised in his somnolent visions on the mid-Atlantic coast. But he also proposed that the secrets of Atlantis would re-emerge in the twenty-first century to enlighten us as to what we should do about it all. That has not yet happened. Most archaeologists argue that there was no such place as Atlantis and that the number of candidates for the lost land around the globe merely show that numerous floods, earthquakes, isostatic shifts and volcanic eruptions have shaped our modern world and will continue to do so. But nevertheless, Atlantis endures in the common imagination, adapted by cultures that inherit the mythos and take it in new directions. And it is not always about how humans have somehow fallen from a state of grace. There are some who derive hope from the notion of a lost undersea world.

Between 1992 and 1997, a Detroit techno outfit under the name Drexciya put out a series of 12-inch records with cryptic clues on the packaging about the submarine origins of the music, including TECHNO FROM THE DEEP and DEEP H2O inscribed into the vinyl. Another record suggested that there was secret knowledge contained within these releases – 'Aquatic knowledge for those who know'. Eventually the liner notes on a compilation of their EPs in 1997 fleshed out the story: that Drexciya was an underwater kingdom populated by the descendants of burdensome pregnant slaves tossed overboard en route to America. These women had given birth to babies who could breathe underwater. They had evolved, deep under the Atlantic, where they built a city. But now they had begun to emerge from the Gulf of Mexico, where they had been sighted, occasionally, as half-human, half-fish creatures in the Louisiana swamps. From there, they migrated up the Mississippi River basin towards the great lakes of Michigan, communicating their message through the medium of techno. 'Do they walk among us?' asked the liner notes. 'Are they more advanced than us and why do they make their strange music?' Through their alternative Atlantis, Drexciya attempted to reconcile the traumas of slavery with the possibilities of transformation, in which the descendants of those slaves would achieve agency in an evolving world, given a new voice via electronic technologies.

This idea was yet another incarnation of Afrofuturism, an aesthetic that emerged in the 1960s and utilised art, fiction and music to speculate on possible futures for the African diaspora. Avant-garde jazz musician Sun Ra – an avid reader of occult literature who had Edgar Cayce's books – posited Atlantis as a forgotten civilisation of Black people who seeded the advanced knowledge that created the pyramids, much as African Americans might yet seed a utopian future in outer space. His space jazz epic 'Atlantis', from the concept album of the same name, begins with pulsing electronic oscillations, like chthonic alarm signals from the depths of the ocean, before erupting into a shattering chaos of discordant clavinet.* Bill Caraher, a professor of history at the University of North Dakota, writes that, for Sun Ra, 'The mystical origins of Egyptian society did not sever contemporary Black culture from an African past, but anchored it in a technologically, intellectually, and spiritually superior civilization that white society had sought to suppress.'[6]

---

* An amplified clavichord.

In 1968, Jimi Hendrix recorded '1983 ... (A Merman I Should Turn to Be)', a sprawling experimental track in which the narrator escapes civil unrest to journey beneath the sea to start his life again in Atlantis, a city brimming with optimism. A decade later, George Clinton's funk band Parliament released *Motor Booty Affair*, a concept album about Atlantis, a place where people were free to dance and express themselves without inhibition. Overton Loyd's album art included a banner with the declaration 'We gotta raise Atlantis to the top'. In his 2006 track 'Underwater', Ghostface Killah's narrator dives into the ocean where he becomes lost among mermaids, Noah's Ark and the wreck of the Titanic. Eventually he beholds a city of blue, gold and white, from a time before Christ, where he is greeted gladly by people clutching copies of the Quran and the Torah, whom he joins in mutual prayer. These musical works reframe the idea of the sunken city as a crucible for the advancement of Black Americans – a place defined by joy and hope, a refuge in time of prejudice and war.

Since its Platonic origins, the Atlantis legend has been a popular vessel for humanity's dreams and nightmares, used to transmit messages across cultural and temporal boundaries about where we came from and where we are going. For Afrofuturists, Atlantis is a place where it is possible for a repressed people to become liberated, transformed into something new. For white supremacists, it is an excuse to colonise cultures they deem inferior and reshape the world in their divinely gifted image. For occultists, Atlantis is a medium for coded spiritual truths. For environmentalists in a time of global warming and mass species extinction, its message is that no civilisation is too great to fall, and that our separation from nature is perilous. For some archaeologists, Atlantis is emblematic of the many ancient sites lost to sea-level rises throughout the early Holocene, and for others it was a real civilisation lost to a flood cataclysm at the end of the Ice Age, which we have collectively forgotten, but for the uncannily similar myths that underlie almost all ancient cultures and religions.

It might seem outlandish that an advanced civilisation could have existed in a deep, unremembered past, with magical technologies not yet discovered. But imagine a version of Edgar Cayce a thousand years from now, after floods and fires have ended our industrial civilisation. Motorways, power plants, factories and server farms are rubble beneath the roots of forests, roamed by deer and dogs. In a woodland clearing, the future Edgar Cayce lies in a hammock by a campfire, over which skewers of rats are being turned. At his side, a group of inquisitors awaits his words with baited breath. He falls into a trance and tells them about our aeroplanes and rockets; our interconnected computers and smartphones; our ability to augment our minds with artificial intelligence and send messages through invisible signals in the air. The amazed listeners, chomping on barbecued rodents, have no idea how these feats could be possible, or how vast reserves of knowledge from the past could vanish from memory. As they ponder this, there's a flash of light in the night sky – a shooting star! They look up, unaware that it is, in fact, a broken panel from an ancient satellite, falling out of orbit.

Meanwhile, in the far reaches of space, the Voyager 1 probe travels through the Oort Cloud with a golden phonograph record containing the secrets of our world, its creators long forgotten, towards interstellar civilisations yet to be born.

6

THE SINKING ISLES

Once upon a time, Lyonesse was a prosperous kingdom in the south-west of England. More than 140 churches hosted pious congregations in towns and villages across an immense forest that ranged along a neck of land in the Atlantic Ocean. The terrain was so flat and low-lying that if you stood on a hill, the distant sea could be seen on either side.

At the centre of Lyonesse was the city of Lions, a place of great chivalry, where knights jousted before jubilant crowds, lords feasted handsomely on the land's great bounty and choirs sang in its majestic cathedral. One of the bravest knights, named Tristan, was heir to the throne. He was sent to Ireland by Meliodas, the King of Lyonesse, to bring back Isolde, the daughter of the Irish king, so that she could become Meliodas's bride. However, during the journey, Tristan and Isolde were tricked into drinking a magic potion and fell deeply in love with each other.

Shortly after they arrived in Lyonesse, their affair was discovered, causing great turmoil in the kingdom. Tristan fled to the court of his uncle, King Mark, in Cornwall. Meanwhile the populace fell into violent recriminations over who was to blame, and whether Tristan could remain the rightful heir. The gods grew angry and whipped up a great tempest in the Atlantic Ocean. High waves crashed on both sides of the land, inundating the forest, surging through villages, churches and farms.

Earlier that evening, a hunter named Trevelyan had fallen asleep beneath a tree on some high ground. He was woken by booming thunder to see the forest below him flooding. In terror, he leapt onto his white horse and rode at a furious speed towards the hills of nearby Cornwall as a tidal wave rose high behind him. Floodwater foamed at the hooves of his steed as they raced to safety.

The sea never receded. On that fateful night, the Kingdom of Lyonesse disappeared from the face of the earth. Over the centuries, its memory faded in the minds of the Cornish people, but for stories told by fishermen of strange relics hauled up in nets, and the ghostly sound of bells drifting across the surf on the Seven Stones reef. The only thing that remained above the sea was an island that had once been the highest westerly peak of Lyonesse, now marooned in the Atlantic, far from the Cornish shore, a place of mysterious tombs and sunken walls, constructed by a forgotten people.[1]

M y memory of the sinking isles is like a dream. We sailed south from the coast of Ireland to the mouth of the Bristol Channel, then beyond Land's End to the big blue expanse of the Atlantic Ocean. Eventually, we spotted a landmass on the horizon, shrouded in mist. Dolphins swam in formation at our bow as we approached, as if escorting us to shore. Towers of granite, topped with grass and golden gorse, rose on each side as we entered a narrow channel between two islands. Behind us, the horizon vanished behind a gossamer veil. The air was pungent with brine. Waves slapped the keel in a funereal rhythm to the drone of the boat's motor. A craggy islet loomed from the swell: a tapered monolith, cracked and bulbous, on a dome of jumbled rocks, lapped by the sea. Drifting closer, I sensed unseen entities shift within its folds and heard a gull's warning cry from somewhere beyond the pale.

On the apex of the isle was a hangman's gibbet and noose, black against the white mist. It passed over us like a threat. I wondered why it was there and what it could mean. It was the era before smartphones so I could not look it up. Instead, I had to accept the object as it appeared, weird and inscrutable. In that moment of wonder, I did not realise that this archipelago was once a single island or that it was linked to the legendary sunken land of Lyonesse. But if I had known, I might have understood why. The isles were so low in the water, they seemed the last vestiges of something greater beneath the surface. Rocky outcrops jutted from the sea like the tips of a myriad Mount Ararats, screaming with sea birds and crowded with seals.

It was 2001 and my friend Duncan had been asked to take his dad's 35-foot boat from the docks in Glasgow to Falmouth. He invited me and three others on the mission. I was a twenty-seven-year-old copywriter living in south London. I had no experience of boats. But he explained that we'd use the engine most of the time so it would be easy. Speed was of the essence, as we had only three days. We packed clothes, beer and food, then took the train to Glasgow. At the end of day one, we moored in Dublin and got drunk in the city. On day two we were hungover and green to the gills, buffeted by a turbulent Irish Sea. We passed over the lost lands of West Wales, then south towards Cornwall. Duncan decided that we would stop at the Isles of Scilly and make our way into the English Channel the next morning.

After Hangman Island we sailed to the largest island, St Mary's, and moored in its harbour. I have since digitally excavated an old outbox, where I discovered an email I sent after the trip, describing how we took the dinghy to a beach where we cooked sausages on a disposable barbecue, then frantically sought a pub before closing time for cigarettes and carry-out beers. But I have no recollection of this. Over two decades of life have become layered on top of the memory, compressing it into a flinty figment. That islet with the gibbet, spectral in the mist, is all that remained.

Now I was returning to the Isles of Scilly, this time on the *Scillionian III* ferry in an unseasonably warm October with the post-school-holiday crowd of pensioners and birdwatchers. The ship pushed away from Penzance, passing St Michael's Mount, an island reachable at low tide by a narrow causeway, with a twelfth-century priory on the top. In Cornish it was known as Karrek Loos yn Koos – 'the grey rock in the woodland' – a linguistic trace of a time when the bay was filled with a forest. Pine- and oak-tree stumps have been exposed on beaches after storm tides, carbon dated between 4,000 and 6,000 years ago, which means they coexisted with the drowned forests on Pett Level, Rhyl and Cardigan Bay. The more I explored the coastline of England and Wales, the more I could sense its ghostly submerged perimeter shelf of forgotten cultures.

I watched the island shrink behind me as the ferry powered into the Celtish Sea, trailing a river of froth and the stench of diesel. Shearwaters flapped towards a distant lighthouse on a spume-battered rock while a gaggle of elderly men on deck pointed high-powered lenses at them, murmuring appreciatively. Then we were out in the queasy swell, the mainland receding into a smear. As I stared down over the side of the ship, I tried to envisage the trees that once grew below, where Stone Age people roamed a landscape that might have been the origin of the mythic Lyonesse.

During the Last Glacial Maximum, the seabed beneath me was a land bridge to what are today known as the Isles of Scilly, the great ice wall stopping just short of its northernmost flank. By 9500 BCE, the meltwaters had transformed Scilly into a single landmass, 17 kilometres long and 8 kilometres wide. But the island continued to sink, thanks to post-glacial isostatic tilt. In the warming climate of the early Holocene, the ocean filled with meltwater and thermally expanded, inundating the lowlands, fragmenting Scilly into the archipelago of 140 islands that exist today. Most are uninhabited outcrops. Some were populated then abandoned. Traces of field boundaries scar the intertidal zone, with walls glimpsed in the shallows. Today's Scillionians live on just five islands: St Mary's, Bryher, St Agnes, St Martin's and Tresco. But as isostatic movement continues, these islands are under acute threat from rising seas and storm waves.

After a couple of hours at sea, I saw the island of St Mary's, with its golden beaches and undulations of grass, bracken and gorse. We docked at the quay in a thin neck of land filled with shops, pubs and houses, hemmed in by the sea on both sides. My lodgings were on the hill above: a hotel in a seventeenth-century military garrison, surrounded by crenelated walls and gun batteries, cannons aiming through palm trees. Passenger boats cruised down glistering blue channels between yellow sand banks. Sun glimmered on green-slimed foreshores. Masses of rocky rubble, inches below the surface, basked in the warmth of the Gulf Stream, awaiting their temporal daily emergence. The elevations of surrounding islands were like the humps of whales, barely breaking the waves. There was nothing around me that was over 50 metres above sea level. The archipelago seemed in the final stages of a geological process that did not have much farther to go.

I took a walk along the island's coastal path, passing white houses on sandy roads, lined with yuccas, pampas grasses and

purple succulents. Abandoned rowing boats lay on scrubby slopes. Mobile phone masts and telegraph wires scored the blue sky. Wooden signs directed me to some of the Bronze Age sites on the island's periphery, including Bant's Carn and Innisidgen burial chambers. Long before these chambers were constructed, Mesolithic visitors felled the island's oak, ash and hazel wood-lands to make dwellings,* despite incremental flooding around them. By 7000 BCE, the island's western parts were fragmenting into islets and the island known today as St Agnes had separated from the mainland. By the time Neolithic settlers introduced grazing animals and permanent settlements, high tides had begun to separate St Martin's and Tresco from St Mary's. As the Bronze Age dawned, 100 square kilometres of land had already vanished. But while the most extreme sea-level rises came before 2500 BCE, it was the relatively small leap between 2500 and 2000 BCE that saw 60 per cent of the land sink, equivalent to an area the size of

---

* Evidence points to 'a definite Mesolithic presence on Scilly but there is not yet enough data available to assess whether the Islands supported a permanent all-year-round hunter-gatherer community'. https://researchframeworks.org/sherf/palaeolithic-and-mesolithic-resource-assessment/

a football pitch each year. Within a few generations, communities became separated by marshes and sand banks.[2] I wonder what the settlers made of such rapid and dramatic changes to their island, stranded in a tumultuous Atlantic Ocean. But whatever they felt, they did not flee. They resettled on higher ground, growing crops, rearing animals and taking advantage of abundant wildfowl and shellfish in the newly formed wetlands. They buried their dead in chambered tombs and built cairns on the headlands.

Throughout the Iron Age, the tilting of the land continued, transforming lowlands into wetlands and further splitting up the landmass. In Roman times, the main island was known as Scillonia Insula, while the Cornish called it Ennor, the 'great island'. But by 500 CE it was an archipelago of fifty-five islands. On the north-west of St Mary's was Halangy Down, a village established in the Iron Age to replace an earlier settlement lost to the sea. Its remains were a series of terraces and slopes, with circular indentations and blocks of stone carved with niches. Archaeologists have identified drains, hearths, seating and food storage, as well as a large house that stood in the centre of the village. Its residents would have seen the sea consume their lands, generation after generation, but they lived fruitfully, farming sheep, pigs, cows and horses, burning bracken for warmth, growing crops and foraging for shellfish. They spun wool, worked iron and made pottery, which they traded with boats coming from Europe and the British mainland.

The apparent success of such communities, even as floodwaters inundated their territories, has inspired climate-change geographers to consider how we might adapt our own society to rising sea levels. In a 2020 study,[3] a research team used the Isles of Scilly as a means to understand the complexities of rising seas, which don't always lead to societal collapse but sometimes to positive adaptation. 'In the past, we saw that coastal reorganisation at the Isles of Scilly led to new resource availability for coastal communities,' says lead author Dr Robert Barnett. 'It is perhaps unlikely that future coastal reorganisation will lead to new resource availability on scales capable of supporting entire communities. More certain, though, is that societal and cultural perspectives from coastal populations will be critical for responding successfully to future climate change.' To illustrate, he points to two examples: Fiji and Tuvalu.

The remote Polynesian island nation of Tuvalu is a circle of atolls and reef islands in the Pacific, averaging 2 metres above water. They are vulnerable to tropical cyclones, but also perigean spring tides, which bring high water when the moon is closest to Earth. A satellite-based system alerts them to extreme weather events, but it might be the dissolution of their environment that ultimately forces their hand. The coral is dead, its bleached remains expelling poisonous algae, contaminating the fish and causing sickness in those who eat them. Saltwater has seeped into the soil, ruining crops and contaminating drinking water. Two islands are almost submerged already. At the UN Climate Summit in 2021, the nation's foreign minister, Simon Kofe, made a video address from a lectern while standing waist-deep in sea water. 'In Tuvalu, we are living the reality of climate change and sea-level rise, as you stand watching me today at COP26,' he said. 'We cannot wait for speeches, when the sea is rising around us all the time. Climate mobility must come to the forefront.'[4]

Faced with a similar threat of rising seas and intensified storms, the island nation of Fiji has established a 'Climate Relocation and Displaced Peoples Trust Fund', the first of its kind in the world. Rather than force people to move, Fiji favours a community-led voluntary approach. On the island of Vanua Levu, residents from the village of Korolevu have resettled on higher land, 2 kilometres inland. And on Viti Levu, the village of Tukuraki was relocated in 2017 after heavy rains triggered a landslide that killed four residents. There will be many more island nations who suffer like this in the coming decades.

According to a World Bank Group report in 2018 a rise in global sea levels is predicted to cause the migration of over 140 million people by 2050. What will happen to those people's memories of the places they left behind? For instance, in the Pacific Ocean's Marshall Islands, there's a rising trend for people to name their children after the atolls, coral heads and islands that could disappear by 2055, when floods are predicted to overwhelm most of the Marshallese terrain and poison their freshwater with salt. By naming their children after important features they might preserve them in the cultural memory of the migrant diaspora. One mother said, 'The names would bear even greater emotional weight if they represented an island that once was but has now ceased to exist.'5

There was a similar story in the San Blas islands off the Caribbean coast of Panama, which I visited in 2003 as I travelled through South and Central America. I boarded a sailing boat in the dock at Cartegena, in Colombia, where a Spanish couple offered me passage to Panama for a fee. They told me we could spend a day at the San Blas Islands, an archipelago of 365 islands, forty-nine of which were inhabited. We moored in the shallows near *Robinson Crusoe* desert islands, circles of sand with palm trees, surrounded by aquamarine water. Minutes later, members of the Guna people rowed out on canoes, laden with lobsters and vibrant handmade fabrics known as molas. After washing the seafood down with tins of Aguila beer, I snorkelled on the reef, marvelling at colourful fish I could not name. Afterwards, I hauled myself onto a sandy island where I sat in the blazing sun. Basking in my European privilege, belly bloated with lobster and Colombian lager, I was blissfully unaware of an imminent ecological tipping point and that these islands would be under threat of submergence within two decades. Children of those Gunas I had just met would be forced to leave for the mainland, while others would build makeshift walls of wood, stones and coral to keep out the sea. Storms are more frequent and destructive. High tides surge through their settlements. A metre of sea rise will put an end to this fragile world and leave the Gunas homeless.

This fate lay in store for St Mary's, too, where I stood atop the hump of the Porth Hellick Bronze Age chamber tomb, looking south over a heathery moor to an inlet that was being eaten away by the tides. Behind it, Porth Hellick pool, the largest body of freshwater on the island, was in danger of contamination, should saltwater flood the dunes. By 2080 the maximum wave height would be a metre higher, threatening to overwhelm its sea defences. Most of the housing stock, infrastructure and businesses were located on low-lying land, with a third of St Mary's less than 5 metres above sea level. The most pressing threat was to Hugh Town, the administrative centre on the sandy isthmus beneath my garrison lodgings, which would be one of the first areas to slip under the waves, leaving my hotel separated on a newborn island.

Shortly before 2021's United Nations COP26 climate talks in Glasgow, the Environment Agency urged the British government to understand that climate-change processes were already under way and sea levels would rise 78 cm by the 2080s, no matter what happened. We had to start factoring these outcomes into the way we managed our urban and rural environments. This included the restoration of peatlands and wetlands along with more green spaces and freshwater storage. 'Adapt or die,' warned the Agency's chair, Emma Howard Boyce.[6]

These measures were already being taken in the Isles of Scilly. Local politicians had declared a climate emergency, urging for safeguards to housing and infrastructure. The Climate Adaptation Scilly project had raised the sand dunes, planting them with protective vegetation. Improved drainage systems removed flood-waters before they reached the fresh water supply. The wetlands on St Mary's were managed through seasonal reed cutting to ensure they didn't dry out, maintaining their role as freshwater catch ment zones and flood defences. One such wetland was behind the inlet of Porth Hellick, a short walk down from the chamber tomb. There I wandered along a boardwalk through swathes of common reed and sedge, like the one in Wicken Fen, lined with information signs about the flora and fauna. It was evident that serious measures were being taken to mitigate against the inevit-able. Archipelagos such as Scilly were brutally exposed on the frontline of global warming and acutely aware of its realities.

To find out more, I took a trip around the eastern isles with Rafe Ward, on his boat *Calypso II*. He was a native Scillionian and, like me, a father of young children. As we sailed along-side St Mary's, watching the remnants of the abandoned Iron Age village pass by, I asked him if he had concerns about global warming. 'If the sea level rises massively, we're going to get really affected here,' he answered. 'At an extreme high tide, the biggest of the year, you see the beaches covered in bits of wood to stop water flooding the streets.' The other problem, he told me, was freshwater. Most of it came from bore holes drilled into natural reservoirs. These were becoming depleted, particularly after hot summers with thousands of visitors filling the hotels and holiday homes. A couple of years ago they reached a worryingly low level. This was why there was such sensitivity to climate change. 'On the whole we are environmentally conscious,' he said, brightly. 'There are lots of grants around to encourage businesses to improve their energy efficiency, and reduce carbon emissions, from buy-ing electric vehicles to installing better windows.'

I sat at the stern as Rafe expertly steered the *Calypso II* away from St Mary's, between tricksy reefs and sandbars, towards a scattering of twelve small islands. The blue sky was bifurcated by a straggly aeroplane vapour trail. Gannets torpedoed the waters. Seals basked on a sandy beach beneath a bristle of granite plates, like the ossified spine of Godzilla. Black-headed gulls amassed on a ledge, wetted by geysers of spray. On the scree slope below, a cormorant fanned its wings. As we skirted the island of Nornour, a flock of shags, bobbing on the sea, dived simultaneously, as if at the behest of an unseen conductor. But while this rugged island gave the impression that we had intruded on an antediluvian non-human world, it was once a populated place, back when a low-lying plain connected it to St Martin's and St Mary's. Those hungry shags I'd seen vanish from the surface were now diving over the ruins of sunken Roman walls.

In 1962, a storm battered the eastern isles of Scilly with gale-force winds and brutal waves. On the island of Nornour it churned up the earth, scattered rocks and revealed the remnants of circular dwellings, like those on Halangy Down. Archaeologists uncovered evidence of an occupation that lasted from the late

Bronze Age until the end of the Roman era. Their finds included 3,000 brooches, rings, bracelets and coins, along with a sword with a decorated scabbard and figurines in the image of the goddess Silina, which was the root of the Roman name for the main island. The evidence suggested that this was once a busy port abandoned to rising waters, which later became the site of a Roman shrine, where traders left offerings in return for safe passage on their voyages from Western Europe to the British mainland. It was once a place of great meaning and importance, imbued with a divine power. But if it hadn't been for that storm in 1962, this history might have been forever forgotten beneath the stones, occasionally stumbled over by a clumsy cormorant, or warmed by the belly of a yawning seal.

The next day I boarded a crammed passenger boat to the island of Bryher in the north-west of the archipelago. The skipper's dog lay on a chair behind his master, eyeing me warily, while a couple of birdwatchers chatted excitedly about the rare snow bunting recently spotted there. On the way we passed Samson, an island formed of two breast-like mounds, scarred with the remains of Bronze Age graves and eighteenth-century cottages. There had been seven households dwelling on the island in the early nineteenth century, subsisting on shellfish and potatoes, until they were removed by the owner for the sake of their nutritional health in 1855. Nobody had lived there since. Our destination, Bryher, had a population of around eighty, only a handful more than Samson at its peak. Those residents lived in a lowland area in the centre of the island. But its sparkling coves, bays, beaches and inlets were under threat from erosion and flooding, with eight sea-defence works proposed at the time of my visit.

I stepped off the boat onto the quay beside a churchyard with lichen-speckled gravestones. A dirt track took me past farms and rustic stone houses, their gardens resplendent with palm trees and succulents. Linnets flitted busily in the hedgerows as a distant tractor engine chuntered in the field and a boy on a rickety bike wobbled past me. It could have been a scene from any time in the past eighty years. But this rural idyll soon fell away and I regressed to a more primal era as I ascended a series of rocky elevations, topped with grass, on the western side of the island. Gulls wheeled in the wind as spray erupted from eroded cliffs, braced against the might of the Atlantic Ocean. But for a few protruding crags there was only the ocean between me and the coast of America. The great stacks of granite were deeply lined, like the wrinkled face of W. H. Auden, wise to the inevitability of death: 'Cold, impossible, ahead / Lifts the mountain's lovely head', as the poet wrote.

At the apex of my ascent, I stood on a protrusion of 290-million-year-old rock, the tip of a magma pluton that extended 10 kilometres down. The word *pluton* comes from Pluto, the Greek god, also known as Hades, whose underworld was the place people travelled to after death. This pluton was part of a batholith that ran all the way to Cornwall and Devon, forming the spine of England's south-west peninsula. It was like the monstrous hulk of a Lovecraftian deity, operating on a chronology that dwarfed my fleeting existence. After all, the story of humans is but a footnote in the evolution of Earth. Its molten flows, tectonic shifts and lithification had been ongoing for 4.5 billion years and would continue for another 7 billion. Standing on the exposed tip of that granite mass, I felt like a bacterium on a whale's skin, unable to comprehend the organism of which I was part, and never to know where in the ocean it was destined – or even why it was travelling at all.

As I tried to imagine the rock that connected Bryher to Cornwall, I mentally stripped away the last 15,000 years. The sea parted, as if a submarine was surfacing, and up from the depths rose the land bridge. Now I was no longer on an island but on the lofty terminus of a peninsula in the aftermath of the Last Glacial Maximum. I saw a periglacial terrain of thawing tundra, gushing with milky outfalls from the receding ice. Then I began fast-forwarding a hundred years a second. A grassland sprouted. Then a forest of hazel, oak and birch, rich with insects, birds and mammals. Soon humans arrived and sat around fires, flaying skin from prey with flint tools. A world thrived before the ground grew marshy, forests sinking into a brackish mulch. Eventually the sea drowned the land bridge, leaving an island 45 kilometres from the coast, emblazoned in the path of the setting sun. It was regularly visited by traders sailing on the Stone Age sea route from Iberia and France to Wales, Ireland and Scotland. Perhaps these people kept alive folk memories of the lost forest between this sacred isle and the mainland, stories that would eventually take the form of what we know today as the legend of Lyonesse.

Since at least the twelfth century, there have been circulated tales of a submerged kingdom between the Isles of Scilly and Cornwall. Perhaps it was the sight of tree stumps around St Michael's Mount, or fishermen bringing up artefacts with their catch, that sparked local imaginations. Or perhaps it was the same root story of Ys and the Lowland Hundred, carried up and down the prehistoric coastal highway, that attached itself to topographical anomalies in the shallows.

Lyonesse was said to be a forested land with 140 churches. At its centre was the cathedral city of Lions. Like Atlantis, it sank after a flood, wiping out every trace of its existence but for the sound of bells tolling from the Seven Stones reef. Its first recorded mention is in the poem *Tristan and Iseult*, a precursor to the Arthurian tales, where it was the birthplace of Tristan, a knight of the round table. Many incarnations of the myth have circulated since then, including versions that tell of a great unnamed crime committed by the people, for which they were punished with a flood. A few hundred years later, the existence of Lyonesse was considered a fact. In a fifteenth-century account, William of Worcester wrote of a lost kingdom between St Michael's Mount and Scilly. The Elizabethan map-maker John Norden referred to 'the supposed drowned lande of Lioness'. It also appeared in William Camden's *Britannia* (1586), which influenced the work of Cornish poet Richard Carew in 1602, suggesting that the Seven Stones reef was a remnant of Lyonesse. The following century, the antiquarian William Borlase noticed walls in the shallows, later speculated to be the remains of Lyonesse by archaeologist O. G. S. Crawford. The legend was popularised in the Victorian era when Alfred Lord Tennyson published a cycle of poems, 'Idylls of the King', in which Lyonesse was the setting for King Arthur's final battle, 'Where fragments of forgotten peoples dwelt / And the long mountains ended in a coast / Of ever-shifting sand'.

Lyonesse continues to attain new significance in modern times. In her electro-pop album *Le Kov*, Gwenno explores the sunken Cornish kingdom. The back cover reads:

> Between Land's End and Scilly rocks
> Sunk lies a city that the ocean mocks
> A bustling metropolis
> Links reaching the furthest seas
> A Cornish capital
> The place of memory.

Her album is sung in Cornish, an almost extinct language with very few fluent speakers. Gwenno links the loss of Lyonesse to

the loss of a language, which takes with it the cultural inheritance of those who spoke it – their names, concepts and symbolic
frameworks. This happens with alarming frequency. Every forty
days, another language dies. This process is exacerbated by global
warming, as indigenous cultures on island nations face inundation from storms and rising seas. Anastasia Riehl, the director
of the Strathy Language Unit at Queen's University in Kingston,
Ontario, says, 'It seems particularly cruel that most of the world's
languages are in parts of the world that are growing inhospitable
to people.' When small communities are forced to migrate and
integrate with bigger societies, their language is no longer spoken
by their children and, quickly, the memory of it begins to fade.
This will happen to half of the 7,000 global languages before the
end of this century. It's another devastating effect of radical environmental change. And it is not an event on the distant horizon
but happening now. As Dougald Hine, co-founder of the *Dark
Mountain* project, writes, 'When we talk about "collapse", there
is a temptation to imagine a mythological event which lies out
somewhere in the future and which will change everything: The
End of The World As We Know It. But worlds are ending all
the time; bodies of knowledge and ways of knowing are passing
into memory, and beyond that into the depths of forgetting.'[7]

In his album *Toll*, Cornwall-raised electronic folk musician
Kemper Norton makes a poetic connection between the Lyonesse
legend and the *Torrey Canyon* disaster of 1967, when a super-
tanker ran aground on the Seven Stones reef, spilling 32 million
gallons of oil where the Cathedral of Lions was supposed to have
stood. The slick coated hundreds of kilometres of coastline and
killed thousands of sea birds. It was drawn south by the current, where it amassed on the coast of Guernsey, an island in
the English Channel. In a desperate attempt to save its beaches,
engineers pumped the oil into a quarry, where it remains today,

emitting a rancid stench, killing birds who land on it, mistaking the black surface for land.

The location of the fabled sunken kingdom Ys was also devastated by an oil spill in 1978, when the *Amoco Cadiz* hit rocks off the coast of Brittany. And in another synchronicity, the *Deepwater Horizon* disaster of 2010 leaked 200 million gallons into the Gulf of Mexico near an oil field named Atlantis, creating a 'dead zone' of oxygen-starved water, with contamination still found in corals, arthropods and fish. In the aftermath, a former BP contractor warned that an oil rig named *Atlantis PQ*, which pumped even more oil than *Deepwater Horizon*, was unsafe and it was only a matter of time before disaster struck again. It's uncanny how events from the mythic past and ecological present can intertwine, which is what Kemper Norton's *Toll* so elegantly expresses. 'The poison drips onto the land, and fills the heart of every man', he sings in 'The Tide'. 'Each night we dream the same old dream, and it keeps returning.'

———

Flood disasters haunt the south-west coast. On 11 November 1099 CE, a monk named Florence of Worcester recorded how the sea 'overflowed the shore, destroying towns and drowning many persons and innumerable oxen and sheep', an event that some think could have been the source of the Lyonesse tale. It happened again in 1755, when a major earthquake struck Lisbon on the coast of Portugal, triggering a tsunami that rolled over a thousand kilometres north to the Isles of Scilly and Cornwall. This remains the most worrying and imminent threat. After a recent sequence of quakes, marine geologists believe that a tectonic plate might be peeling apart beneath the seabed, creating a new subduction zone, where one plate begins to sink under another.[8] A major earthquake event is likely, with a 10 per cent chance of a 'potentially damaging' tsunami in the Gulf of Cadiz in the next fifty years. Begona Perez, Head of the Division of Oceanography of Spanish Ports, says, 'The question is not whether there will be another tsunami, but when will it happen.'[9] In such an event, Neil Hamlyn of the Devon and Cornwall Local Resilience Forum has warned that the Isles of Scilly 'could be covered and disappear'.

In such an event, the last elevation in the isles to vanish would be the precise point on Bryher where I walked on a carpet of heather, away from the headland, along the eastern flank of the island. Known as Watch Hill, it is the highest spot in the archipelago, from which lookouts signalled to pilot boats when sailing ships approached. On the other side of a narrow waterway was the island of Tresco, its coastline running parallel. Rounding the bluff of Watch Hill, I spotted an islet in the middle of the channel – a protrusion of bare granite with small boats moored nearby. It took me a few moments to realise that this was the islet I had encountered when I came to Scilly twenty years earlier and the same sea corridor between two islands I had sailed down with my friends. It was discombobulating. In my head, I knew it was Hangman Island. After all, there was a gibbet on top of it. But it looked nothing like I remembered. The object in my mind's eye was more isolated, the rock blacker, its environs barren and empty of people. Whereas, in reality, it was yards from a sandy beach in a flowery cove with houses and a cafe.

Our pasts are largely fictions. We each navigate through life in a whirl of subjective thoughts, emotions and sensory phenomena. Much of it registers on the subconscious level. When we recall places, and what happened in them, it's like trying to remember a dream. The haphazard fragments fade after we wake, and words are inadequate to express such a complex experience. Some details fall to the wayside. Others become more prominent. When we tell the story, it's an approximation. We fill in gaps and exaggerate elements that enhance the narrative. As years pass, we forget that those embellishments never really occurred, for we are no longer accessing the physical place but the story we told about it. This is why places are more than the sum of their topographical parts. They are also figments of the imagination; shifting memoryscapes that alter in meaning, slippery and unreliable projections of the mind. Hangman Island was such a place for me. The islet was objectively and measurably the same, yet it had mutated in my memory, becoming hyper-real and mythic.

I thought of everything that had happened since I first encountered this rock as a young man: my expedition through Latin America, my marriage, the births of my children, the books I had written, my divorce, new friends I'd made and old friends I'd lost. Then I thought of the world events that had unfolded: 9/11, war in the Middle East, Brexit, the pandemic. Hangman Island felt like the fulcrum on which the world had turned from one state into another. Whatever picture I had had about the future in 2001 was gone. Something strange and nightmarish had taken its place. Fires burning in Arctic tundra. Microplastics in our bloodstreams. Mass extinctions. Cryptocurrencies. QAnon conspiracies. Artificial intelligence and algorithmic social media reality tunnels. This otherworld of 'now' would have seemed alien and frightening to that younger me. Equally, that prelapsarian world in which I sailed carefree with my friends on a boat to the Isles of Scilly was a fiction. My memory of Hangman Island has little in common with what really happened. I don't recall much about that trip, nor about my adventure through the San Blas Islands on my way to Panama. Indeed, most of my life since then has been irretrievably forgotten.

I sometimes yearn for that past through which I have travelled but can never return, nor even accurately describe, and which exists only as a series of fragmentary memories. But perhaps that is all 'the past' really is. According to some of the latest theories in quantum physics, time doesn't objectively exist; rather it's an emergent property of our minds, a tool through which we organise information about experienced events. 'We understand dreams as a mental construct,' writes the American scientist Robert Lanza, 'but when it comes to the life we live, we accept our perception of time and space as absolutely real.'[10] But, according to Lanza, time and space are not real outside our minds. Instead, time is 'a mode of interpretation and under-standing. It is part of an animal's mental software that moulds sensations into multidimensional objects.' If this is the case, then our memories *are* the past, and therefore that past – as well as the places we passed through and the objects we saw – are inherently mental constructs, subject to inaccuracies, embellishments and erasures. I console myself that while memories are unreliable and forgotten events are unrecoverable, they are not meaning-less. Each of us is an aggregation of feelings, thoughts and ideas, gathered up like glacial moraine. It makes us who we are, even if we don't know where it all came from.

The same goes for human history, 98 per cent of which has never been recorded. Cultures have risen and fallen. Languages have died. Major planetary events are missing from the collective memory. Important ports, cities and temples submerged. Islands vanished. But these lost things still exist. Everything that has ever happened is contained in the present. Humans are an accumula-tion of inherited experiences going back to our ancestors on the plains of Africa. We are each a living history. If only we could remember ourselves.

When I got home from my trip, I attempted a linguistic approximation of what I thought and felt seeing Hangman Island for the second time. But, even as I type these words, the memory is already like a fading dream. It will be edited by my publishers, who will send it to bookshops for readers to interpret. My experience will enter the written record, where it will remain so long as there are copies in the world. It might continue to exist after my death, gathering dust on a shelf, to be read one day by someone not yet born. They will assemble their version of me based on the assumptions of their inherited cultural framework. Perhaps they will tell a friend about it in their own words, and the listener will conjure the scenario in their imaginations, and then tell it to another friend, creating the shoots of a new folklore.

They might wonder what the Isles of Scilly looked like in that time before the floods, just as today we might wonder about Lyonesse. For sea levels have risen and Hangman Island is now beneath the waves, with only its gibbet revealed at low tide. One day, a boat approaches with five friends on board. They spot a worn, seaweed-strewn stump sticking up above the surface like an ancient relic.

Puzzled, they wonder what it is, why it is there, and what it could mean.

7

INSIDE THE VOLCANO

Over 2,000 years ago, Baiae was a town in the Bay of Naples where the greatest leaders of the Roman Empire came to play. The waters that ran from the hills were miraculously hot and soothing. Healing steam rose from the earth. To take advantage of these natural wonders, a terraced network of bathhouses lined the slopes, filled with holidaymakers, who gorged on local oysters and wine. Soon, Baiae became regarded as a place of sin and decadence, notorious for drunkenness, infidelity and debauchery, where the party never stopped.

One afternoon, a mighty explosion brought the revellers rushing onto their terraces in various states of undress, half drunk and dripping with water. They watched in horror as, on the opposite side of the bay, a column of cloud blasted up from Mount Vesuvius like a tall pine tree, spread out at the top like branches, before it descended in a thick black rain of stone.

A darkness fell across the land and the waves rose high and boisterous. The ground shook violently and the sea seemed to roll back on itself, driven from its banks by the convulsive motion of the earth. Fish and dolphins were left stranded on the empty beach, flapping and writhing, before the waters gushed back again with force, crashing to shore and surging through the low-lying villas.

For a moment the people of Baiae wondered if they had incurred the wrath of the gods for their sinful ways and were doomed to destruction. But they survived the eruption and it was those poor souls in Pompeii and Herculaneum who perished in a poisonous deluge of pumice. Perhaps Baiae had been saved by the gods for a greater destiny, being a haven of rulers, leaders and thinkers.

But the shattered mountain on the skyline was not the only volcano in the bay. The holidaymakers did not know it, but they themselves were perched over a gigantic chamber of magma and gas. As this began to empty out in the dying years of the Roman Empire, their beloved resort slowly sank.

Eventually, the waters took all memories of this place down into the depths of the bay, their roads, villas, shops and statues covered in sand.

Then, in the twentieth century, the lost land began to rise again . . .[1]

The broken cone of Mount Vesuvius haunts the tower blocks of Naples in a haze of pollution. Three million people in the city go about their business while a magma chamber boils beneath them. They are aware of this threat, unlike their Roman ancestors, who farmed the volcano's fertile slopes oblivious to its secrets. When it erupted in 79 CE it destroyed the city of Pompeii in a deluge of toxic gas and pulverised pumice. But Vesuvius has a larger and more dangerous sibling.

Twenty minutes' drive west along the Bay of Naples is the Phlegraean Fields, the caldera of a super volcano with a 15- by

12-kilometre crater, most of which lies beneath the waters of the bay. Above the surface is a ragged arc of craters, cinder cones and sulphurous vents. Villages are dotted among its hills. Narrow roads wind through terraces of lemon, fig and olive trees. On the western shore of the bay are the ruins of Baiae (today known as Baia), a holiday resort for the Roman Empire's elites. Julius Caesar, Mark Antony, Nero and Hadrian owned villas there. It was a place where the rich and powerful could let their hair down. Emperors, senators, poets and commanders frolicked in a luxurious complex of thermally heated bathhouses overlooking the harbour. Sipping wine in their robes on balconies, lovers by their sides, they gazed at Vesuvius on the skyline with no idea that their beloved resort was destined to sink.

Baia is the closest thing we have to an explorable Atlantis – a debaucherous city of advanced architecture and great political influence destroyed by volcanic activity and covered over by the sea. Its villas and bathhouses lie in the shallow waters of the bay, offering me a tantalising opportunity to dive down and touch its submerged ruins. It is located right in the heart of the Phlegraean Fields, a constantly shifting volcanic landscape that has sunk most of Baia below sea level since Roman times. But unlike similarly depressed places like the English Fens, this sunken land is rising again, thanks to the expansion of subterranean molten rock and gas beneath Earth's crust. A lost Roman world is returning to the water's surface, offering us a chance to gaze on its wonders, before it might be taken down once more, through either steady collapse as the magma and gas leach away again or – as some scientists fear – a major volcanic eruption that could destroy the land in a single catastrophe and potentially change the planet forever.

The Phlegraean Fields is steeped in myth and legend: it is the site of the entrance to the Greco-Roman underworld; the home of ancient Greek oracles; and the origin of sacred prophecies about natural disasters. Only 3 kilometres north-west of Baia are the ruins of Cumae, the first Greek colony established in Italy. Previously, it had been settled by Bronze Age people with an oracular tradition, described in Homer's *Odyssey* as guardians of the entrance to the underworld. Prophetesses, known as sibyls, dwelled in Cumae's caves, where they foretold the future for a fee. One day, a sibyl emerged into the light with nine papyrus books and commenced the 200-kilometre trek north to Palatine Hill.

In Rome, she demanded to speak to the king, Lucius Tarquinius Superbus, who reigned in the final years of the sixth century BCE. He met her in a courtyard at the edge of the city, where she warmed herself by a bonfire. The sibyl explained that in her possession were books of immense value to the king and his descendants. They contained prophecies of monumental events, along with rituals that would mitigate against any disaster. Would the king be interested? Tarquin said he might be, depending on the price. But when she told him that price, he laughed in her face. Not on your life, he said. The sibyl shrugged, drew three books from her bag and tossed them onto the fire. 'Each refusal will lead to the annihilation of further prophecies and to the premature end of your city,' she said. Thereupon she offered him the remaining six books for the same price. This was extortion, thought the king, shaking his head.

Three more books burst into flame, illuminating the sibyl's face in the glare. The king stared with horror into the depths of her pupils, all the way down to the molten bowels of Earth where rock spewed and sparks flew. In that instant he knew she was the real deal. Perhaps he had been too hasty. After all, the future of his kingdom was at stake. As the sibyl prepared to consign the remaining prophecies to the fire, Tarquin shouted, 'I'll pay!' He bade his attendant fetch a bag of coins from the royal hoard. Once the cash was in her hands, the sibyl vanished, leaving three intact books, their pages flapping in the breeze beside the embers of the dying fire.

From that day forth, those tomes were prized possessions. After Tarquin's death, they were stored in a stone chest inside the Temple of Jupiter, where they remained under armed guard. Whenever there was a natural disaster or conflict, the books were summoned to be consulted by the Senate of the new Republic to reveal what rites they'd need to appease the angry deities. During a drought in 241 BCE, the sibylline prophecies advised that the Romans should build a temple to Flora, the goddess of flowers and fertility. As war raged against Hannibal later that century, they offered a solution in the form of a sacrifice, demanding that two Greeks and two Gauls should be buried alive. They didn't, however, prophesise that the Temple of Jupiter would burn down in 83 BCE, along with the books themselves. But so important were they that the Romans sent out emissaries to gather as many sibylline prophecies as they could find that closely matched those which had been destroyed. These new editions were kept in the Temple of Apollo.

In 15 CE, heavy rains swelled the Tiber beyond its bursting point and waters flowed through the city, submerging buildings, drowning citizens and animals. For many citizens, this was a sign that they had most egregiously offended the gods. The Roman senator Asinius Gallus urged Emperor Tiberius to consult the sibylline prophecies so that they might avert further catastrophe. But the emperor was resistant. He believed that referring to supernaturally divined prophecies would undermine his authority, which was why the devious Asinius Gallus was pushing so hard for it. Ancient Greek priestesses were no match for Tiberius's rationality and logic. He would deal with the crisis in a practical way. Instead of sacrificing a Greek or building yet another temple, he ordered a commission of five senators to look at new flood-control measures. According to the Roman historian Dio, these were designed 'so that it should neither overflow in winter nor fail in summer, but should maintain as even a flow as possible all the time'.[2] But this did little to prevent similar incidents in the centuries to come.

Ancient Rome was no stranger to floods. One was even featured in the empire's origin myth, starring the famous twins Romulus and Remus. The story goes that a jealous noble named Amulius deposed his brother, King Numitor, and took his crown. He exiled Numitor and murdered his sons. In a moment of clemency, Amulius spared the daughter, Rhea, but ordered her to remain a virgin so that no offspring could exact their revenge. However, Rhea became pregnant, claiming that she was raped by Mars, the god of war, and gave birth to twin boys. Rather than kill them outright and risk the wrath of Mars, the king ordered the babies, Romulus and Remus, to be abandoned in the Tiber. The river was flooding when the king's steward tried to cast the babies' cradle into the waters, so they could not be forced into the main current, which would have ended their lives in drowning or starvation many kilometres away. Instead, he placed it in the river's periphery, where it drifted downstream to the base of Palatine Hill, where stagnant pools formed in the marshy floodplain. There, the basket was laid down gently by the receding waters.

A wolf discovered the babies and nursed them as if they were her own. Then a local shepherd took them under his protection, raising them to adulthood. When the twins eventually discovered their true identities, they helped overthrow the evil King Amulius. Afterwards, they decided to create a city on the hill where they had been saved by the wolf. But they clashed over who would be ruler. Romulus murdered Remus and it was his name given to the new city: Rome. It became the capital of an empire ruled by mighty Caesars, who continued Romulus's tradition of plotting and backstabbing so that they could cling to power, or take it from others.

Romulus's city was founded on seven hills, but construction spread across the Tiber's drained marshes, while the forests on the slopes around them were cut down at a rapid rate, leaving Rome vulnerable to inundation. There were thirty-three recorded major floods between 414 BCE and 398 CE, but the historian Professor Gregory S. Aldrete believes that the majority went unreported and might have occurred every couple of decades.[3] Roman historians described farmhouses, theatres and bridges reduced to ruins, statues toppled, coffins dislodged from graves, devastated crops and famines. In one account from 23 BCE, the flooded city was navigable by boat for three days. In another from 69 CE, the Tiber swelled to such a size that it flooded higher parts of the city.[4] The Roman historian Tacitus later described how 'many people were carried off in the open;more were trapped in shops and in their beds'.[5]

Remarkably, while Romans had the engineering expertise to control the flow of the river, few plans to do so were carried out. This may have been down to the general infrastructural robustness of the city* or a fear of punishment from the gods for meddling with the divine Tiber and thwarting their acts of deluge. It's worth understanding this when we try to imagine what the Romans might have felt about the loss of their notorious coastal party town, Baia. Mostly, they interpreted floods as prophecies and omens that might help guide their political decisions and overseas war campaigns, not as an indictment of their urban-planning policies. The dominant philosophy of the empire

---

* Aldrete points out that Rome's innovative network of drainage sewers and aqueducts for safe drinking water meant that it quickly recovered from major floods. Its structures of brick and marble were resilient to water damage and its hilly topography provided its citizens with safe havens.

was stoicism, in which the violent vicissitudes of nature – and the terrible losses of life that can result – were to be accepted with wisdom, courage and moral fortitude. Everything happened for a reason, even if we didn't understand it.

Just as in ancient Rome, we have constructed substantial amounts of our housing on floodplains in the modern era.* In the past decade one in ten of new homes in England has been built on high-risk areas.[6] There's a similar situation in the USA, where over 14 million properties are located in the hundred-year flood zone.[7] 'The rains came down and the floods came up,' goes the old song and yet we continue to build our houses on the sand.

This became startlingly apparent the week I was due to fly to Naples, when Storm Eunice pounded Britain and northern Europe with 196-kilometre-per-hour winds. Falling trees crushed cars. Trucks were overturned. Roofs shredded. Homes plunged into blackness. Barely had the debris settled when meteorologists warned that another storm would make landfall the day before my flight. Three hundred British flood warnings were in place. An evacuation centre was set up in a Manchester mosque in case the River Mersey broke its banks. In Doncaster, the River Don topped its embankments, unleashing torrents of fast-flowing water. While floodgates at Avonmouth, Lydney and Sharpness on the River Severn resisted the tidal surge that swept along the estuary, fourteen people had to be rescued from Ironbridge in Shropshire when the river burst its banks and houses slipped underwater.

In the stormy days before my departure, I nervously watched the carnage on my smartphone: torrents of dirty water coursing through streets, the Millennium Dome roof in tatters and

---

* Almost 1.9 million homes have been built on floodplains according to figures released by the National Audit Office in 2020.

aeroplanes struggling to land in shear winds. My anxiety went into overdrive. All my life, I've been prone to compulsive catastrophising. My mind habitually locks into worst-case scenarios. The sound of a distant explosion? Nuclear bomb. A call from my dad at an unusual time of day? Dead relative. Persistent headache? Probably a brain tumour. On a cliff walk, I'm beleaguered by visions of falling over the edge. Driving a car on the motorway I see crashes unfold before me, one after another. Aeroplanes are crucibles for these nightmarish visions. Throughout a flight I repeatedly see the fuselage crack open, fire roaring into my face, burning bodies plummeting.

Climate breakdown added a new layer of mental discomfort. A return flight from London to Naples released 250 kg of CO_2 per passenger into the atmosphere, double the amount that one person in Gambia produces in a year, a country suffering the brunt of climate change as seas flood its rivers with saltwater, causing crop failure. As a young man I had moved through the world with a clear conscience, enjoying the thrill of travel without guilt. But no longer could I pretend I was not culpable. Global warming tainted almost every action, from switching on my computer, to shopping at the supermarket, to opening my fridge, to firing up the gas hob. Each contributed in some way to fossil-fuel extraction, deforestation and CO_2 emissions. I couldn't turn on a TV, read a newspaper or look at social media without seeing news of record-breaking heatwaves, freak floods, collapsing ice shelves, forest fires or cyclones. I was beleaguered daily by graphic visions of a doomed world collapsing in a mayhem of extinction, fire and flood. Perhaps this was what it was like being a sibylline prophet in Cumae, besieged by portents of disastrous futures. Try as I might, I could not stop these thoughts. Instead, I had to carry out everyday tasks while they continued like a silent horror reel in a cordoned-off section of my mind.

On my knuckle-whitening ride through turbulent skies over the European Alps it occurred to me that scuba diving for the first time, stress hormones flowing through my veins, days after a flight through the immediate aftermath of a storm, might not be my greatest idea. But I emerged with relief from the airport to find the weather calm and sunny. From Naples I drove along a series of unfolding elevations around the curve of the bay, until I reached my hotel on a hill overlooking Baia. From my balcony I looked across palm trees and lemon groves to the harbour where waves danced towards a dock lined with restaurants. On a slope behind the town were the terraces where emperors once languished in geothermally heated baths. On the other side of the harbour a promontory was topped with a medieval Aragonese castle, built on the site of a Roman villa suspected to have been owned by Julius Caesar. The bucolic vista, bathed in Mediterranean sunshine, was exhilarating. It felt like a safe refuge from the chaos back home, even if this landscape was actually the surface of an active super volcano that could erupt at any time and turn my bones to vapour.

I descended a winding trail through rustic country houses, where cats slept in shady gardens among piles of lemons. As the path twisted steeply down towards the sea I noticed the remnants of Roman walls, made from diamond-shaped blocks and crude concrete. At the bottom, what appeared to be a short cliff was a partly submerged Roman building, lapped by waves. It was my first view of Baia's sunken ruins.

A little further on was the Temple of Diana, a concrete dome, like half an Easter egg on its side, built in the time of Emperor Hadrian. It was made of tuff – porous rock formed from volcanic ash. Entrances to vaults peered from banks of grass, suggesting parts of the structure that had sunk into the earth. The term 'temple' was misleading because this building was far from

religious. It contained thermally heated baths and – some have speculated – a casino.[8] In its prime it would have been decorated in vibrant colours, filled with artworks and statues, resounding with conversation and laughter. Now its grey carcass looked over a contemporary plaza ringed with fake colonnades and plastic faux-Roman urns filled with sand for people to stub out their cigarettes. To one side was an abandoned local information office, glass cracked, posters peeling from the walls. The ruins of twenty-first-century tourism in the ruins of a third-century tourist destination.

Modern Baia consisted of a single road lined with shops, cafes and restaurants below the balconies of residential flats, draped with drying clothes. There was no bank. No cash machines. No supermarket. The main reason visitors came here was to board yachts or tour the underwater archaeological park. Diving centres displayed photos of the submerged ruins, showing me the first glimpses of the larger part of the ancient town that lay beneath the water.

I sat outside a dockside restaurant, looking across the bay to Pozzuoli, once known as Puteoli, a major Roman city in the Phlegraean Fields, where ships unloaded cargoes of grain from Egypt and outgoing vessels were loaded with volcanic sand, an ingredient of Roman concrete, for export to construction projects around the Mediterranean. Nearby was Lake Avernus, a flooded crater, which the Romans used as a secret naval base. A kilometre-long passage allowed horses and chariots to travel underground from the base to Cumae. In a terraced complex on the hill behind where I sat, Roman generals and their consorts would take time off in Baia's famous thermal baths and 'sweating rooms' where hot vapour rose from subterranean springs. The site was excavated in the 1950s by Amedeo Maiuri, the archaeologist who oversaw digs at Pompeii and Herculaneum. His team hacked into a vine-covered hillside to reveal bathhouses, spa rooms, swimming pools and lodgings for guests and servants, linked by ramps, steps and corridors.

After lunch I explored what remained of this complex above the water, walking its avenues of truncated colonnades and broken walls. Some remnants of stucco, mosaic flooring and wall art were intact, portraying water nymphs, dolphins and athletes. At ground level was the Temple of Mercury, also known as the Echo Temple, an enormous flooded concrete dome. Built in the first century CE, its pioneering design would influence the construction of the Pantheon in Rome. Baia was a place of architectural innovation, a testing zone for projects in the capital.

Like many mythic sunken kingdoms, Baia was notorious for its moral corruption. By all accounts, it was a wild place.

Adultery was rife. Drunken holidaymakers caroused on the beach after dark, skinny dipping and singing rude songs into the wee small hours. Roman groupies were invited onto the boats of wealthy patricians for late-night booze cruises. The scholar Marcus Terentius Varro wrote that 'nubile young women were common property, old men behaved like young boys, and a lot of boys as though they were girls'. The poet Ovid called it 'the appropriate place to make love' while, in a poem addressed to his lover Cynthia, Propertius (elegiac poet and friend of Virgil) expressed alarm that she was intending to travel to the resort, where men would certainly attempt to seduce her. It concluded:

> Just leave corrupt Baiae as soon as possible.
> Those shores will bring divorce to many,
> shores unfriendly to chaste girls.
> Go to hell, waters of Baiae, you crime against love![9]

The Stoic philosopher Lucius Annaeus Seneca also held a dim view of Baia. 'I left it the day after I reached it,' he wrote in one of his letters. 'Baiae is a place to be avoided, because, though it has certain natural advantages, luxury has claimed it for her own exclusive resort.'[10] For Seneca, Baia was a 'village of vice', ridden with behaviour that was 'foreign to good morals'.

By the fourth century, the town was in a steep economic and social decline as the Roman Empire crumbled. It was also physically sinking. Beneath the Phlegraean Fields, the magma chamber is constantly emptying and refilling, pushing the terrestrial surface up and down, like the chest of a sleeping giant, in a phenomenon known as bradyseism. In the dying years of the empire, the land was lowering, as gas and lava leaked away, and the Mediterranean was pouring in. This was around the same time the Romans in the English Fens were cursed to a watery grave by the Iceni people's sea god. Perhaps this was the same furious deity, taking down the decadent party HQ of Rome's hubristic elite.

Year by year, Baia's villas, shops, thermal baths, oyster farms and roads slipped beneath the water. Mosaic floors were covered in a blanket of sand. Statues, marbles and frescoes encased in silt. Molluscs colonialised its broken walls. Crabs sheltered beneath broken vases. By the eighth century CE, more than half of the town had sunk. What remained was raided and sacked, first by the Barbarians, then by the Arabs. It was finally abandoned in the year 1500 after outbreaks of malaria. But the town was never quite lost. In his seventeenth-century travel guide, Maximilien Misson wrote, 'One can see many ruins of temples, baths and palaces; and some of these remains are visible under the surface of the sea. The vicinities of the city were dotted with pleasure villas. Today, these are nothing more than grim remnants that turn this once enchanted place into a place of ghastly desolation.' These ruins would later capture the imagination of Romantic poet Shelley, whose 'Ode to the West Wind' describes how the wind wakens the Mediterranean Sea:

> Lull'd by the coil of his crystalline streams,
> Beside a pumice isle in Baiae's bay,
> And saw in sleep old palaces and towers
> Quivering within the wave's intenser day . . .

The sunken city glimpsed in the water expresses Shelley's idea that fallen civilisations give rise to new ones. Much as spring follows winter, disaster can birth hope. In his poem, the west wind brings about revolution in the same way that a poet can facilitate a transformation to a better society. In his final lines

Shelley implores the wind to scatter his words like sparks from a fire to wake up the people of Earth so that they might take action and create the change that is necessary for their liberation from the mistakes of the past.

—

North of Baia is Lake Avernus, the crater lake that was supposedly the entrance to Hades. It features in the sixth book of *The Aeneid*, a Roman Empire creation story written by the poet Virgil, set in the volcanic world of the Phlegraean Fields. It describes how the Trojans arrive at Cumae and meet the sibyl. Their leader, Aeneas, asks her to predict if they will successfully settle in the region of Latium (the future location of Rome) to the north. She tells him they will meet with war and misfortune, but there will be a path to safety. Aeneas requests entry to Hades so that he might speak about this with his dead father, Anchise. The sibyl dutifully guides him to the entrance, which is found beneath the volcanic Lake Avernus. From there, Aeneas journeys along the River Styx through zones of slain soldiers, sinners and pure-hearted souls until he finds his father, who confirms that their descendant Romulus will one day establish the city of Rome, and a man named Caesar will bring about its Golden Age, when its empire will extend far overseas.

The notion of a world beneath a volcano also forms the basis of Jules Verne's 1864 novel, *Journey to the Centre of the Earth*. When I was a boy, fascinated by lost prehistoric worlds, I avidly watched the movie version, starring Kenneth More. He plays Professor Otto Lidenbrock, who leads a team of adventurers into an Icelandic volcano that's the entrance to a weird subterranean sea with islands populated by giant monsters. This was a direct influence on my story, *The Travel to the Underwater Palace*, which I wrote when I was seven. My protagonists' journey by boat through islands filled with volcanos, dinosaurs and demons, with most of the sailors killed along the way. After the survivors reach the looming spire of the underwater palace, they don their diving suits and plunge overboard. At the doorway to the palace, they are met by those companions previously devoured by dinosaurs or murdered by cavemen. The underwater palace turns out to be an underworld where the living and the dead can co-exist, just as Virgil described in *The Aeneid*.

Keen to see the entrance to Hades for myself and reignite my boyhood fascinations, I took the short walk from Baia to the shore of Lake Avernus, a large disc of choppy blue water, fringed by bulrushes. The surrounding hills were low and stepped with vineyards, slopes of golden tuff rising above them, peppered with shrubs. This pretty idyl was a far cry from what the crater lake used to be in the Bronze and Iron Ages, when sulphur escaped from vents in the lakebed, poisoning wildlife and emanating a foul stench. The name Avernus comes from the Greek term *a-ornis*, which means 'birdless'. But today it was bobbing with dozens of ghost-faced coots. At the lake's eastern end was the shattered ruin of a majestic Roman bathhouse, partly lost to the waters thanks to the sinking effect of bradyseism and the carnage caused by the nearby hill, Monte Neuvo, when it erupted from the earth in 1538.

It was easy to find the sibyl's cave without any Indiana Jones shenanigans because it was a pinned location on Google Maps. I walked a single-lane road along the south bank of the lake to a dilapidated wall with faded capital letters that read: 'GROTTO DELLA SIBILLA'. A trail led through a mulchy woodland to a descending pathway, sided with brick. At its terminus was a vertical wall of rock, straggly with dead vines. A flight of stone steps led to a rusted iron-grilled entrance, secured shut with red chain locks. I shone my phone torch into the dark interior. There was a tipped-over wooden chair, with four stones placed in a cross and a witch's broom. I couldn't work out whether this was an accidental agglomeration of objects, an arranged set piece designed to unsettle visitors, or the aftermath of an occult ceremony.

Most archaeologists believe that this grotto was a Roman construction for military purposes, built during the time when the lake was a naval base, and not the lair of a psychopomp at the entrance to hell. A more likely location was in nearby Cumae where, among the many ruins, was the Temple to Apollo. It was built in the sixth century BCE, before the reign of King Tarquinius Superbus, when this was a Greek colony. In the *Aeneid*, Virgil writes that the temple was founded by Daedalus, father of the legendary Icarus, who was famous for flying too close to the sun. Near this site, claimed Virgil, was the cavern of the sybil, she who 'reveals things yet to be'. In 1932, a long trapezoid corridor was excavated by Amedeo Maiuri, the same archaeologist who would uncover the bathhouse complex in Baia. There were multiple entrances along its length, fitting Virgil's description of the cavern, including a chamber where the prophetess might have lived.

Most intriguing was the theory that the entrance to the underworld was in Baia itself. During his excavation, Amedeo Maiuri discovered a narrow tunnel network dug into the hill behind the ruined complex, plunging 50 metres down from the entrance and leading to an inner sanctuary. It was claimed that toxic gases were leaking from inside, and so the entrance was closed off. However, in the 1960s this hidden feature caught the attention of amateur archaeologist Ferrand 'Doc' Paget, who had retired to the area. He theorised that the tunnel was deliberately engineered as a facsimile of the entrance to Hades by Romans seeking to capitalise on the myth. He braved the risk of asphyxiation and probed deep into the interior. The tunnel led to an underground stream that could easily have doubled as the River

Styx. He posited that visitors were taken across a narrow water-way to the inner sanctum to meet a fake sibyl, whom they'd pay to hear their fortune. Professional archaeologists have argued that it's more likely that the construction of interlinked tunnels was designed to channel hot vapours to the facilities at the surface.

Given the Phlegraean Fields' abundance of Stygian tunnels, subterranean streams and sulphurous vents, it's unsurprising that fact and fantasy so seamlessly flow into each other, or that it inspired legends of hellmouths, supernatural prophetesses and chthonic deities, like Vulcan, the Roman god, who dwelled inside the Solfatara crater. The volcanic topography is the stuff of dreams and nightmares. Toxic fumes. Birds dropping dead over lakes. Mud that strips flesh from bones. Ground that quakes. Molten rock spewing from the bowels of the earth, killing veg-etation and contorting the landscape into alien shapes. A major eruption could instantly wipe out local ecosystems and soci-eties. Traumatic memories of such events might have become enshrined in mythology. For instance, the historian Mott T. Greene believes that the eruption of Thera, the volcano in the Aegean that tore apart the island of Santorini, inspired the Greek legend of the Titanomachy, where Olympians fought the Titans in an epic clash.

Certainly, the ancients were no strangers to seismic cata-clysms. In 426 BCE, earthquakes triggered a tsunami in the Aegean, devastating islands, leaving the town of Orobiae under water. Another earthquake off the coast of Crete in 365 BCE sent a wall of water rushing down the Nile delta towards the Egyptian city of Alexandria. The Roman historian Ammianus Marcellinus described how waves levelled houses 'and the whole face of the world seemed turned upside-down'. Older even than Alexandria was Thonis-Heracleion, an Egyptian port city that was much like modern Venice, a network of canals, buildings and harbours on interconnected islands in the Nile delta. In around 150 BCE, an earthquake struck, toppling a colossal statue of Hapi, god of fertility and guardian of the river, into the waters. The ground shook so violently that the waterlogged soil turned to liquid and the buildings sank as a tidal wave crashed over them. While some remnants of the city were inhabited during the late Roman era, by the eighth century it had vanished beneath the Mediterranean.

A similar fate befell the Greek city of Helike in the Gulf of Corinth in 373 BCE, when an earthquake liquified the soil before a tsunami sank what was left. Its citizens worshipped Poseidon Helikonios, god of earthquakes and seas, so their annihilation was deemed punishment from their patron deity. It was such a widely discussed disaster among Greek historians that Plato would have heard about it when he wrote about Atlantis. The city didn't completely vanish, though. A century and a half later, the mathematician Eratosthenes spoke with boatmen who claimed that a bronze statue of Poseidon in the Gulf of Corinth got entangled in their nets. In 174 CE, the Greek geographer Pausanias saw ruins in the shallows of an inland lagoon. Afterwards, Roman tourists visited the site, just as I had come to Baia to view the spectacle of a sunken city. Over the centuries, tonnes of silt were dumped into this lagoon by feeder rivers and the city was presumed lost forever. That was until 1988, when archaeologists began searching for Helike, using sonar to scan the seabed. When they didn't find anything, they moved inland, where they discovered walls, roads and pottery alongside seashells under 3 metres of sediment. The silt-covered bones of Helike had returned to the surface a kilometre in from the sea.

It was a lifting of the land that brought the ruins of Baia back into the modern consciousness. After centuries of decline, the magma chamber beneath the Phlegraean Fields began to refill, forcing them up towards the surface again. In the 1940s a pilot named Raimondo Baucher took aerial shots that revealed their haggard outlines in the bay's calm waters. Since the 1960s, underwater archaeologists have uncovered the perimeter walls of thermal baths, fish farms and shops; mosaic floors; and a cobbled road that once led from Baia to Portus Julius in Puteoli. They also discovered the remnants of a grand villa owned by Emperors Claudius and Nero. In its grounds was a nymphaeum, a rectangular building dedicated to water spirits, which would have been a place of respite for drinking wine and feasting beside a swimming pool. A collection of its marble statues was embedded in the sand, including two believed to represent Dionysus as a child, and one showing the face of Antonia Minor, Emperor Claudius' mother. These statues were brought to the surface to be restored and replaced with replicas.

In 2002 the Archaeological Marine Park of Baia opened the ruins to the public. To view them I had two choices: snorkel or scuba dive. It seemed the latter would give me my best opportunity to touch real sunken ruins. For the past two years I'd gazed with yearning over dark British coastal waters for that which was unseen – and in some cases what might never have existed at all. But now I would get the sense of what it might be like to uncover a lost city beneath the waves, and live out that fantasy I had aged seven, when I wrote *The Travel to the Underwater Palace*.

In the office of the diving centre, I was handed a form filled with small print, absolving the company of any responsibility should I not declare any extant health problems. These included a dizzying array of symptoms, most of which I believed I had. My throat was feeling sore. There was a weird buzzing sensation in my chest and I was short of breath. I also noted that diving was not advised for anyone with hypertension, something I had become convinced I suffered from but could not know for sure because the very sight of a blood-pressure monitor sent my blood pressure soaring. I signed the legal document anyway. Like my fictional sailors, I had come too far not to visit the underwater palace. But unlike those heroes I imagined as a kid, I was a neurotic middle-aged man with no diving skills.

A young, handsome diving instructor with clearly excellent blood pressure strapped air tanks and weight belts to me as the motorboat cruised towards the large Roman ruin jutting into the sea. The gear was heavier than I expected, pressing me down in my seat. After the boat anchored, I toppled backwards into the clear water, as instructed. For ten minutes I was shown how to eject water from my mask without drowning, and which signals to use for communication. Thumbs up meant 'go to surface immediately', rather than 'everything is okay', which would take getting used to. Safety checks done, my belt was deflated and down we went.

Beneath me, the sandy bottom of the bay seemed unremarkable. There was no vision of a lost city, no turrets draped in seaweed, no trinkets gleaming in drifts of silt, no underwater palace. But as we pushed forward, I discerned humps of Roman wall and blocks of stone; the hint of human design in the angular geometry of what seemed at first to be rocks. Admittedly, it was hard to focus on the ruins with the sound of my breath rasping in the mouthpiece, blood thumping in my ears and the terror that my heart was about to rupture. But whenever the instructor checked on me, I gave him the okay sign and we flippered across the seabed, towards a jumble of ruins where luminescent fish circled replicas of statues in Nero and Claudius's villa, positioned where the originals had been found. The instructor encouraged me to touch the face of Claudius's mother but it seemed odd to do so, because it was a modern simulacrum installed as a visual aid to enhance the archaeological experience. Then again, what did it matter? Perhaps one day this replica would itself be a historical artefact, discovered by divers of the future, after brady-seism has taken Baia's ruins back into the depths. By then, the rising sea might have inundated the modern town, including the abandoned tourist office on the plaza beside the Temple of Diana. Its fake classical columns and plastic ashtray urns, poking from the sand, would reveal as much about our era as the marble versions revealed about the Roman Empire. A civilisation that thought it would last forever went as far as it could go, then inevitably retracted, as sure as an inbreath leads to an outbreath, or a magma chamber empties after it fills. The future diver might caress the fake urn with fascination and imagine what led a society with such command of petrochemicals to its watery doom.

After I touched the statue, I made a clumsy turn to follow the diving instructor and my flipper struck the top of Claudius's mother's head, wobbling it on its plinth and sending a terrified fish hurtling out from behind it. Eyes wide with panic, the instructor lunged to steady the pretend relic. I didn't know what the sign language was for 'Sorry, I'm an idiot', so I shrugged and followed him to the next location, swimming over carved stone blocks, crude concrete walls and arched brick alcoves, wavy with kelp and flitting with brightly coloured fish.

I tried to conceive these ruins as a luxurious holiday resort, decked out with marble statues and frescoes, where people in togas reclined by a pool drinking wine, flirting with each other under a beating hot sun. It was like floating over the remains of Las Vegas, New Orleans, Mykonos or Ibiza, the notorious 'anything goes' party destinations of our time, reduced to murky fragments in a blue gloom as the keels of boats full of wealthy holidaymakers sliced the surface above. My guide led me down to the seabed, where I could see an intricately patterned black-and-white mosaic Roman floor. These were tiles on which a sandalled emperor likely walked, or a servant knelt to scrub away wine stains and vomit. Because we formed the rearguard of today's tour, my guide's responsibility was to scoop the sand and gravel back onto the tiles to keep them protected from algae. So we both knelt on the floor, like my imaginary servant, and concealed the tiles in the way murderers might hurriedly bury a body, before swimming over the perimeter of the sunken villa towards the boat.

With some relief, I broke through the surface into the sunshine, lifted my mask and removed the rubber mouthpiece. In the distance I could see Mount Vesuvius, grey and pale. Beneath me, its volcanic twin was swelling with molten rock and lethal gas. The bay was pregnant with diabolical potential.

Less than a month after my dive, an earthquake with a magnitude of 3.6 on the Richter scale shook the Phlegraean Fields. Research has shown that the crust above the super volcano is being stretched to breaking point, making it more likely to rupture, unleashing hell from below. Over 600 earthquakes were recorded in this location in April 2023 alone.[11] When the first eruption happened here 39,000 years ago, creating this caldera, it blasted so much debris into the atmosphere that it cast Europe into a volcanic winter, darkening the skies, turning the sun blood red and altering the climate as far away as Russia, Asia and North America. Some vulcanologists have even proposed that this Stone Age cataclysm was a tipping point for the extinction of the Neanderthals.[12]

Volcanoes might have contributed to the end of the Romans, too. Recent research has shown that there was a period of climate instability from 250 to 550 CE, marked by droughts, floods and extreme weather, culminating in a series of volcanic eruptions in the sixth century, more violent than anything experienced by humans for thousands of years.[13] These eruptions triggered what is known as the 'Late Antique Little Ice Age', when temperatures plummeted for the next 150 years. Some historians believe those centuries of climate chaos to be one of the contributing factors to the end of the empire. 'Like any large civilisation – including the civilisation we have today – it was highly dependent on predictability of natural resources,' says Michael Mann, professor of meteorology at Penn State. 'It was very heavily adapted to the climate conditions that had persisted for centuries.'[14] Seismologists warn that a major eruption on a scale unseen since before the last Ice Age could happen again in the Phlegraean Fields, tipping our modern culture into chaos, but nothing is certain in the land of the volcano, except for uncertainty itself.

During their empire, Romans were typically stoic about earthquakes, eruptions and floods. These were elemental forces, unleashed by gods, that not even they could control. 'Think too how many whole cities have "died" – Helike, Pompeii, Herculaneum, innumerable others,' wrote the philosophical Emperor Marcus Aurelius. 'One man follows a friend's funeral and is then laid out himself, then another follows him – and all in a brief space of time.' As a child he spent summers in Baia, writing essays and learning the art of oration. Later, as a troubled emperor, not much older than me, he wrote about how change was inherent to nature, and loss an integral part of change. Therefore it was not to be feared or mourned. Time was a turbulent stream. As soon as one thing came into sight, it was swept away, just as great cities had been destroyed by disasters. 'All things fade and quickly turn to myth,' he wrote. Little did he know that his childhood holiday resort would join his list of fallen cities, or that the Roman Empire itself would crumble to dust, its ruins scattered across its former European strongholds, gawped at by tourists.

It isn't easy being an endemically anxious person in the twenty-first century. Existential perils have piled up faster than my mind can comprehend. Global warming tipping points. Nuclear proliferation. Pandemics. Asteroids. Any or none of these could be our undoing. Sometimes I wonder why the planet isn't full of screaming people, clinging to what remains of their sanity like the safety bar on a fairground waltzer as we whirl around the sun on a rock that is getting hotter. Even if, like the Romans, we try to appease whatever deity is behind all this with an ancient Cumaean ritual, my hunch is that it would have little regard for human fate. So I must learn how to remain stoic in the face of potential annihilation, whether I am sitting on a plane, waiting for test results in the doctor's surgery, watching wildfires turn skies red on the TV or swimming over the ruins of a dead empire in the caldera of an active volcano.

Floating towards the diving boat in my scuba gear, wheezing like an asthmatic manatee, I considered the magnificent vista around me. The devastating eruption that formed the caldera of the Phlegraean Fields, killing everything around it for many kilometres, also created fertile slopes for vineyards, warm shallows for oyster farming and those healing hot springs around which the cities of Baia and Puteoli flourished. Its crater lakes and underground streams inspired myths and poetic epics. Its sands became the concrete that built the cities of an empire. One day the volcano might explode again, but the land it destroyed would be superseded by something new, gifting a terrain to the people who inherited it thousands of years down the line, or to whatever strange beasts crawl over the earth when we are gone.

8

HURRICANE CITY

A long time ago, a spirit named Aba appeared before a North American people known as the Choctaw. He gave them an urgent instruction to build a boat made from cypress wood. They were to fill it with all the local animals, along with enough food to last them for a week. 'It must be constructed on that peak over there,' said Aba, pointing to the highest hill in the land.

Passing strangers were bemused by the sight of the Choctaw cutting, hammering and chiselling all day and night on the hilltop. 'Why build a boat so far from water?' they wondered with amusement.

After the work was complete, they searched the forests and swamps for a male and female of each beast and took them to the boat, where they all huddled together to wait for whatever was coming. That day, winds began to howl as a hurricane brewed. Black clouds billowed and flashes of lightning struck the land. Roofs were torn from dwellings and a hard rain came down, turning the ground to mud. Soon floodwaters poured through the land, forming gullies, which turned into fast-flowing rivers, which soon became an inland sea that expanded over the whole land, rolling with high waves.

On the fifth day, the storm abated. The survivors found them-selves adrift in calm water, surrounded by the floating corpses of people and animals. Apart from the birds in the sky and the selected species in the boat, life on Earth had become extinct.

Finally, the waters subsided and the boat settled on the shore of an island where a single willow tree grew. They cut it down and rubbed the branches together to kindle a fire. Nearby they found a strange white grain, which they planted in the ground to create the first ever cornfield.

And so the world began anew.[1]

The caged boat drifted through the Louisiana swamp. Slowly, we passed the submerged trunks of bald cypress trees, their branches hung with Spanish moss, gnarly roots protruding from the water like skeleton knees. The riverbank was leafy with hackberry, water tupelo and Chinese tallow. Racoons gathered to watch us, wary of predatory eyes that peered from a float-ing mat of rotten leaves below them. Sitting among a gaggle of tourists, resting my forehead against the protective metal bars, I watched the alligators slither alongside, armoured backs glinting wetly in the sunshine. One lay motionless on a log, legs dangling, as a soft-shelled turtle floated through fronds of emerald-green swamp grass like a lost leather baseball glove. Behind the wheel was our guide, Captain Danny, a portly old man who spoke into a microphone with a deep southern drawl. Every now and then, he hurled marshmallows into the river to lure alligators closer for our entertainment. I wondered how their digestive systems, evolved to eat raw meat, coped with processed blobs of sugar, corn syrup and gelatine. When he wasn't baiting 'gators, he told stories about the thirty-five species of snake in the Manchac swamp, the fat bullfrogs he liked to eat, and the ghosts that haunted this cursed place, where a town had been destroyed by wind and flood.

After a bend in the waterway, the boat passed an enclave where an alligator sunned itself on a bank. In front of an iron railing were five white wooden crosses with alligator skulls leaned against them. Beneath each cross, explained Captain Danny, were multiple bodies, victims of a hurricane, in which 200-kilometre-per-hour winds and a 4-metre tidal wave flattened the logging town of Frenier over the course of one devastating day in 1915. A hundred yards downriver, we passed a second burial ground with only one cross, on which was nailed a human vertebra. Animal bones jutted from its apex and spherical bells hung on each side. This was the grave of Julia Brown, said Captain Danny, the voodoo priestess blamed for the disaster.

It's hard to pinpoint where any story truly begins but a good place to start this tale would be where others in this book have begun: 20,000 years ago, when great sheets of ice extended over much of the North American landmass. After the Ice Age ended, meltwaters coursed down the Mississippi River valley, carrying massive volumes of rock debris and sediment, gouging its way south. By 5000 BCE, the river snaked through southern Louisiana to the Gulf of Mexico, where it left silty deposits, creating what would become one of the largest river deltas in the world. When the mountain snows in the north melted each spring, the river flooded, leaving more alluvium to replace that which had settled and compressed, or been eroded by the tides. Layers upon layers built up, creating barrier islands, estuaries and new lobes of sand, silt and clay as the river shifted course in its eternal quest to reach the sea.

One of Louisiana's native peoples, the Chitimacha, made their home in the delta. In their oral tradition, the original world was

a flooded place, empty of life. Its creator, the Great Spirit, made crawfish to live in the waters. He instructed them to dive down and bring mud up to the surface. As they did so, they created the land, just as the annual flooding of the Mississippi River created a fertile delta for the Chitimacha. Further up the valley, the Biloxi had their own flood myth in which the waters rose so high that everyone drowned apart from a woman and her two children, who had climbed a tall tree. Another Louisiana people, the Choctaw, tell a story about how their ancestors escaped a deluge after a warning from a benevolent spirit. Cataclysmic mega-floods were at the foundation of many Native American creation mythologies. But these delta-dwelling peoples were also attuned to the recurring cycles of flood in their own time, delivered by the mighty waterway that constantly reshaped their territory, producing as much as it took away.

A crescent of natural levee on a bend in the Mississippi River, formed from deposited sediment, became a sea trading post for the Native Americans. Behind it was a brackish estuary named Okwata – or 'wide water'. Locals used this bay as a shortcut from the Mississippi into the Gulf of Mexico, via an adjacent lake, Maurepas, which was surrounded by a cypress forest. The Choctaw gave this land the name Manchac, a word for 'rear entrance'. In 1682, the French stole Louisiana from the Native Americans and constructed a settlement on that crescent of high ground, naming it New Orleans. They renamed the estuary Lake Pontchartrain. The nascent city needed supplies and encouraged migrant farmers to make the most of the alluvial soils. In the late 1700s a group of German immigrants settled in the Manchac swamp on the western shore of Lake Pontchartrain. They began logging the cypress forest for its highly prized wood, impervious to rot and termite infestation. They used swamp water to irrigate fields of cabbages, which they turned into sauerkraut. Their settlement, Schlösser – later renamed Frenier – was isolated until a railroad connected them directly to New Orleans, and to the rest of the USA, which had purchased Louisiana in 1803. Now trains carried the town's logs and sauerkraut off to market. But as operations expanded, the removal of the cypress trees eroded their wetlands and made them more vulnerable to floods.

In the late nineteenth century, an African American woman named Julia Brown moved from New Orleans to Frenier with her husband. Local legend has it that she was a voodoo priestess. This religion was brought to the Americas on slave ships from Benin in Africa. After the Haitian slave revolt of 1791, many freed people migrated to New Orleans, where their version of voodoo merged with Catholicism to create a distinct Louisiana brand, dominated by voodoo queens. These women became powerful spiritual and political figures in the nineteenth century, offering healing, counsel and advice. Julia Brown would have been an asset in a secluded town such as Frenier, where it was hard to access medical care. She became a local healer, offering natural remedies for pain and fever, and was so endeared to the locals that they called her Aunt Julia. But after the death of her husband, she became resentful, believing her services were being taken for granted. She sat on her porch by the edge of the swamp, playing her guitar and singing songs with cranky lyrics that hinted at terrible things happening to the residents of Frenier. In the final weeks of her life, one song had this refrain: 'Oh, when I die, I'll take the whole town with me.'

On the day that Julia Brown passed away, the sky grew dark and there was an eerie stillness in the air. If the newspapers had arrived in Fernier that day, instead of languishing in a railway depot in New Orleans, awaiting transit, people might have realised that a ferocious weather front was approaching. But instead, they grew fearful that Brown had brought a curse on them. On the morning of her funeral, 29 September 1915, the waters of Lake Pontchartrain rose rapidly and a strong wind whipped up. As the casket was carried towards the grave, the deluge began. The people ran for cover as rain hammered down. A gale bent the trees and peeled the rafters from the houses. Many townsfolk hid in the railroad depot, which collapsed in the wind, killing twenty-five of them. A wave rolled along Lake Pontchartrain and struck the town, defenceless on the lowland, no longer protected by cypress trees. The Burg family – Joseph, Dora and their children – took refuge in a schoolhouse behind the railroad tracks. They frantically clambered onto a table as floodwaters surged through the building. One of their children, a young man named Martin, scrambled out of a broken window, his little sister Frieda clinging to him, and dived into the churning brine. Frieda lost her grip and was swept away but Martin clung to a tree, listening in horror to the cries of his loved ones. It was the following morning before the waters subsided enough for him to clamber down to safety. By that point, almost everyone had drowned or been crushed beneath falling timber.

After the storm, there were 300 dead in Louisiana, sixty of them in Frenier and the neighbouring settlement, Rudduck. Both places had been wiped out. Corpses lay in piles of debris. Others vanished into the alligator-infested depths of the Manchac swamp. It was said that, from time to time, a skeleton would emerge from the mire. Those bodies that were recovered were laid in a mass grave. Julia was interred separately, her cross adorned with bells to warn trespassers that this was an evil place and they should stay away. According to folklore, the ghosts of the drowned still roam the swamp, their screams audible from the cemetery at night, and the voice of Julia Brown forever sings her prophetic song of doom.

The real story of the voodoo priestess and the cursed town is uncertain. In some retellings Julia Brown is known as Julia White or Julia Black. A New Orleans *Times-Picayune* report on the storm refers to her funeral as being that of 'an old negress who was well known in that section, and was a big property owner', with no mention of her religious status. It may have been that she was a voodoo healer. Such women were often resented for the threat they represented to white power in Louisiana. To undermine it, the voodoo religion was mocked and castigated as savage and evil. Portrayals of voodoo in art, literature and cinema persistently indulged in stereotypes, insinuating that this was a sinister religion based on fear, blood sacrifice and black magic, rather than one that promoted healing, spiritual development and respect for our ancestors. Equally, Brown might simply have been a literate, Black female landowner – another reason for white society to fear her.

Some years after the tragedy, a local sightseeing business placed railings and crosses in the cemetery to mark the spot, including a wooden sign with '1915' engraved on it and a collection of alligator skulls. It could be that the annihilation of Frenier created a terrible absence that locals were compelled to fill through this act of memorialisation; a way of fixing the story to the place so that it would endure. Or perhaps everything I'd seen on the boat trip was simply a prop to help tour operators tell an appealing horror story, using ghoulish, theatrical scenery to manifest the tragedy in a semi-fictional narrative of their own making, where a terrified populace suffered the wrath of an evil voodoo queen. It goes to show how a flood can become unmoored from its factual foundations, gathering supernatural elements and socio-political meanings as it drifts through the generations.

Despite having no shared root with the Celtish myths of sunken kingdoms like Lyonesse, Ys and the Lowland Hundred, the folkloric telling of Frenier's destruction has much in common with them: a town suffers vengeance from the spirits for perceived crimes and slights; it is covered in waves over the course of a single day; ghost sounds reverberate from the lost place many years later. As similar flood disasters must have wiped out settlements like this in the Stone, Bronze and Iron Ages, it is easy to see how their experiences might also become transformed in the oral record, taking on a narrative life of their own. The event becomes a story; then the story becomes the event, retold by the likes of Captain Danny for paying tourists on his boat, meandering down the river of time.

When another hurricane struck this part of southern Louisiana ninety years later, it would also attain mythic qualities, sharing characteristics with many of the fabled ancient disasters in this book. The great city of New Orleans, notorious for its supposed decadence and sin, was protected from the sea by embankments, walls and gates, which its arrogant leaders believed would keep them safe. But it was catastrophically flooded after a storm and more than a thousand of its citizens were drowned. In this case, the city survived but the land on which it sits is sinking faster than anywhere else in the world, while the seas keep on rising. New Orleans could soon become a contemporary Atlantis, submerged before our eyes in the twenty-first century. And it would not be a freak event, but the first among many inundations of coastal cities around the globe.

In the week before my flight to New Orleans, the deadliest storm since Katrina struck Florida's coastal towns, with a wave 4.5 metres high turning streets into rivers, cutting power lines, and leaving the cities of Fort Myers and Port Charlotte in tatters. The storm surge penetrated kilometres inland, flowing into homes, sewers, water supplies and septic tanks, contaminating the water, unleashing flesh-eating water-borne bacteria.* When I booked this trip, I hadn't considered that it would still be hurricane season in the Gulf of Mexico. Mercifully, the skies were clear as my plane started its descent. Gazing out of the window as we turned on the wing, I saw the freeway on stilts that led through the Manchac swamp and a line of electricity pylons on

---

* In the aftermath, contamination from the toxic slurry unleashed the flesh-eating bacteria *Vibrio vulnificus*, https://www.wired.com/story/hurricane-ian-flood-flesh-eating-bacteria-vibrio/.

plinths across Lake Pontchartrain. Finally, I beheld New Orleans: a bright dazzle of lights in a gargantuan black mirror. It was like arriving at a mythical floating city.

Most of what I could see below me as the plane descended was reclaimed marshland. It would have been verdant with cypress groves and rivers full of alligators and turtles when the French first started building houses here. The location was strategically ideal, being a seaport far from English and Spanish settlements, which gave them control of the Mississippi River valley. A nearby bayou led north to Lake Pontchartrain, offering easier access to the Gulf of Mexico than the 160-kilometre-long journey to the mouth of the river. After a hurricane in 1722 flattened the original settlement, it was rebuilt in the grid formation that still exists today as the French Quarter. The natural levees were reinforced to prevent seasonal flooding and New Orleans spread along the crescent of high ground in the swamp, known today as 'the sliver by the river'. This is how urban development progressed after the Spanish took control in 1762 and when the USA purchased it in 1803. By that time New Orleans had become an international hub for slaves, brought on ships from Africa to work on the sugar plantations during the 'white gold' boom. But the Mississippi continued to flood, as it had done for millennia, causing havoc in the city, forcing the authorities to keep digging more drainage ditches and canals, while further raising the levees. Meanwhile the swamp behind the crescent city remained a no-go area for developers hungry to expand on the city's burgeoning wealth.

In the 1910s, a local engineer named Albert Baldwin Wood built innovative mechanical pumps that forced water through canals via a series of interlocking pumping stations. This allowed New Orleans to spread westward and northward into the wetlands. After the 1915 hurricane destroyed Frenier, the authorities proclaimed triumphantly that their drainage system had mitigated against the worst effects. It defences were tested again during the Great Mississippi Flood of 1927, which inundated 70,000 square kilometres of delta in Arkansas, Mississippi and Louisiana with up to 9 metres of water, killing over 500 people and displacing tens of thousands of Black Americans. Hearing news of the devastation upriver, the New Orleans authorities dynamited the levees south of the city, intending to sacrifice the less populated rural area of St Bernard should the flood overwhelm it. But because other levees broke further north, this never happened and New Orleans was saved, for the time being.

By this point the reclamation of the marshland was in full swing. Mechanical drainage pumps were hard at work, drawing water and sludge from the lowland bowl and pushing it out into Lake Borgne and Lake Pontchartrain. Loggers moved in and felled the cypress groves to clear the way for new suburbs. But in the same way that logging had left Frenier vulnerable in 1915, the removal of these tree groves took away the city's natural protection against storm surges and, without the replenishment of seasonal floodwaters, the soil crumbled and shrank, compacted by the weight of roads and buildings. Immediately, the suburbs began to subside. Within fifty years, the neighbourhood of Gentilly sank by a foot and Lakeview by 1.2 metres. By 1960, more than half the population lived below sea level. By now, New Orleans was a vibrant cultural hub, famed for its Mardi Gras festival, jazz music and permissive attitude, which earned it the nickname the 'Big Easy'. But it had also become an oil city, thanks to bounteous reserves of fossil fuels beneath the surrounding wetlands and in the Gulf of Mexico. Thousands of exploratory and developmental wells had been bored into Louisiana's southernmost parishes. Canals were dredged to allow ships, rigs and equipment to move through the marshes, creating conduits for saltwater that killed off freshwater fauna and flora. One of these was the industrial canal, built between Lake Pontchartrain and the Mississippi River, which cut through the Ninth Ward, a densely populated urban area. To connect it to the Gulf of Mexico, the Army Corps gouged a 120-kilometre-long shipping lane through the eastern wetlands, known as the Mississippi River–Gulf Outlet.

When Hurricane Betsy made landfall in September 1965, a 3-metre wave coursed into Lake Pontchartrain, as it had done in 1915 during the funeral of Julia Brown. This time, it travelled down the industrial canal, bursting its levees and flooding the lower Ninth Ward to its east, an area with a largely Black population. The water reached so high that people drowned in their homes. Over 160,000 properties flooded and it took ten days for the waters to recede.* Rumours abounded that the levee had been bombed so that the impoverished Black area would take the brunt of the flood and not the more affluent white area to the west of the canal.

After surveying the carnage, the governor of Louisiana, John McKeithen, promised to put in place 'procedures that will someday in the future make a repeat of this disaster impossible'. The Army Corps drew up plans for an ambitious new system of flood protection. They would build a protective wall around the city, with a 3-metre-tall barrier across Lake Pontchartrain to prevent surges entering the bay, with gates for ships that could close in the event of a hurricane. But over the decades, these plans became compromised. The Army Corps opted for merely heightening and reinforcing the levees, citing cost-benefit ratios as their rationale. These reduced measures did not account for

---

* In the immediate aftermath, Louisiana Senator Russell Long phoned President Lyndon Johnson and said, 'Aside from the Great Lakes, the biggest lake in America is Lake Pontchartrain. It is now drained dry. That Hurricane Betsy picked the lake up and put it inside New Orleans.'

the environmental factors that were causing southern Louisiana to sink faster than any place in the world: sea-level rise, wetland erosion, resource extraction and land shrinkage caused by the very measures supposed to protect it. And so the stage was set for the biggest flood of them all.

On Sunday 28 August 2005, a tropical cyclone with winds of 250 kilometres per hour approached Louisiana from the Gulf of Mexico. The next day, Hurricane Katrina made landfall in Plaquemines Parish, a populated wetland south-east of New Orleans, and then veered north-east, missing New Orleans by 35 kilometres, and continued towards the cities of Gulfport and Biloxi. But just as the city breathed a collective sigh of relief that Katrina had not hit them directly, the storm surge struck. A giant wave rolled into Lake Pontchartrain, while another flowed up the Mississippi River–Gulf Outlet, sending a great swell down the Industrial Canal. The pressure was too much for the inadequate levees, which failed right across the city. Defensive walls collapsed on the 17th Street Canal, flooding the neighbourhood of Lakeview. Walls along the London Avenue Canal also fell, flooding Gentilly. A levee in the St Bernard Parish, made from sand and shells, completely disintegrated, while another protecting New Orleans East from Lake Borgne was washed away. Over a million people had been evacuated but 130,000 remained in the city, 10,000 of whom were huddled in the Superdome.

For two days, water poured into the city, filling the bowl of reclaimed swamp where twentieth-century developments had been constructed. Eighty per cent of New Orleans vanished under water and 1,392 people died.

Cassandra Snyder had come to the city in 1988 to study at Loyola University and became reborn as a New Orleans citizen. She worked in hotels and produced music events, eventually founding the private tour company Soul of Nola. Cassandra had been evacuated before Hurricane Katrina filled her house with a metre of water. She knew first hand the devastation but had been one of those who returned to a city that some had suggested was a lost cause. To show me where the defences had failed so badly in 2005, she drove me along the river, concealed from view by a high wall, to the Industrial Canal, a channel of brown water lined with grassy levees, where a bascule bridge carried traffic behind a steel lock. On the rise of the levee I looked beyond a discarded mattress to an abandoned naval base: graffitied six-storey high-rises behind barbed-wire fences, populated by the homeless until they were recently forced out. 'CAPITALISM IS THE VIRUS' proclaimed red letters along the length of one building. It was on this high ground that survivors from the nearby Lower Ninth Ward gathered after their homes sank.

At 7.45 a.m. on that fateful morning, residents heard an explosion as the canal's floodwall collapsed. Many of them were still lying in bed and there was little time to escape. Water cascaded through homes and shops, carrying away trucks and cars. A 200-metre-long industrial barge crashed through the canal wall, cutting a swathe through several blocks of houses. The floodwaters quickly rose past first floors, then second floors, until they lapped at the eaves, forcing panicked residents to scramble into attics and onto roofs. Yet again, rumours abounded that the walls of the Industrial Canal had been blown up deliberately to flood their neighbourhood rather than white areas. After all, the authorities had done this during the flood of 1927 to sacrifice a parish, so why would they baulk at doing it again?

Author Andy Horowitz has uncovered evidence that, while the walls were not dynamited, there was an official plan to 'trap as much of the flood as possible in the Lower Ninth' via a siphon under the canal. But in his book *Katrina: A History, 1915–2015*, he explains that the main reasons behind the mass failure of the

city's levees were poor design, unfinished works, lack of mainte-
nance and a failure to incorporate knowledge about land sinkage,
wetland loss and sea-level rise. A case of 'lethal arrogance', as one
engineer put it. The floodwalls along the Industrial Canal had not
been anchored deep enough, which is why water undermined the
concrete. The Katrina flood was therefore not a natural disaster
but one created by humans, who had failed to adapt their city
to the realities of the changing environment around it, putting
short-term economics ahead of long-term sustainability.

Cassandra drove me across the Claiborne Avenue Bridge and
then down, over a metre below sea level, to the Lower Ninth
Ward, where a monolithic concrete wall stretched towards a giant
electricity pylon. This was the replacement for the wall that had
failed in 2005 – a last bastion of protection against a future flood.
Turn after turn, on cracked asphalt streets, we passed expanses
of grass, gravel and concrete where homes had once stood. The
new houses built since Katrina – wooden cabins with porches,
raised aloft on stilts – only accentuated those absences in the
spaces between. Former gardens were overgrown with wild

shrubs. Grass sprouted from sidewalks. A mildewed coach was filled with broken wood. We passed a discarded sofa and a dented car, tyres flat, doors flung open. In some of the empty lots, stone blocks showed the footprint of the former building, with steps leading to front doors long gone. But in other areas there were signs of development. Smart houses with glossy white porches and stairs leading down onto cropped lawns. Playgrounds for kids. Experimental new buildings on stilts in an array of shapes and colours, designed to withstand hurricanes and floods better, with solar panels on their roofs.

Unlike other areas of New Orleans, the Lower Ninth Ward has not returned to normality. The Black Americans who first settled here had been allowed to transfer home ownership down through the generations. However, many of the purchases were made through unofficial financial mechanisms, making the details hard to trace.* The grants offered by the 'Road Home' rebuilding scheme required legal proof that they owned their properties. Some couldn't produce it or had lost crucial paperwork in the flood. For most victims, the cost of repair turned out to be more than the market value of the property. With so much damage done to their neighbourhood and no amenities such as grocers, hairdressers and cafes, it was not an alluring place to return to. The population was only a third of what it was before the flood. Across the city, Katrina had indiscriminately destroyed the homes of white and Black, rich and poor. But the recovery exposed deep-rooted inequalities, just as the initial reporting on the disaster had been racially skewed.

In the immediate aftermath of the flood, the media relayed surreal images of traffic lights poking above the water, canoes paddling past roofs and bodies floating in backyards. Dark

---

* Andy Horowitz, *Katrina: A History, 1915–2015*, p. 153.

fictions took hold. Reports announced that armed gangs prowled the city, taking advantage of the chaos to rape, murder and steal. White people were usually portrayed as seeking water, dry clothing and food in a crisis, while people of colour were routinely described as looters. There were no more murders in the week after Katrina than there had been in an average week before the storm. Yet the story being told to the world was that the city's inhabitants had descended into a state of primal savagery. From the Superdome, there were hysterical reports of violent beatings, drug binges and babies being raped, like something out of an eighteenth-century Swiftian satire.

What was really happening had nothing to do with the failings of those people huddled in fear inside the building, but a colossal failure of preparation by city officials. Winds had torn the roof off because the structure had not been adequately tested, even though it had been designated as an emergency refuge. The electricity failed. Air conditioning and refrigeration stopped working. Food went rotten. Temperatures soared and the humidity became stifling. There were inadequate supplies of water. Without functioning toilet facilities, the stadium filled with the reek of faeces and urine. It was a hell, but not of the people's making. Their morals had not been corrupted. Their suffering was not *deserved*. Yet many religious conservatives believed that an unrighteous populace had brought this disaster on themselves, much in the way that the Lowland Hundred and Ys myths blamed lust and drunkenness for their inundation.

Catholic priest Gerhard Wagner described Katrina as 'divine retribution' for New Orleans' permissive sexual attitudes and tolerance of homosexuality. The Reverend Franklin Graham said, 'This is one wicked city, OK? It's known for Mardi Gras, for Satan worship. It's known for sex perversion. It's known for every type of drugs and alcohol and the orgies and all of these things that go on down there in New Orleans.' Michael Marcavage, head of Repent America, an anti-gay Christian group, said, 'The citizens of New Orleans tolerated and welcomed the wickedness in their city for so long.' As the evangelical Christian Zionist Hal Lindsey put it, 'The judgement of America has begun.'

Over three centuries previously, a similar moral framing had been applied to the sinking of Port Royal. This English stronghold on the south-eastern coast of Jamaica was a notorious party hotspot for pirates and plantation owners, reputed to be 'the most wicked and sinful city in the world' with its brothels, gambling dens and heavy drinking culture. The English satirist Edward Ward called it 'the very Sodom of the Universe'.[2] On the morning of 7 June 1692, an earthquake measuring 7.5 on the Richter scale shook the city, liquifying the sandy soil beneath it, pulling its houses, churches and military forts down into the churning earth. The tsunami that followed destroyed whatever else remained. Fifty acres of the city disappeared under water, with over 2,000 dead.

Barely had the aftershocks finished when the earthquake was described as divine punishment for the city's sins. The Reverend Emmanuel Heath, a recently appointed rector of Port Royal, described his congregation as a 'most Ungodly Debauched People' and declared the catastrophe to be a 'terrible judgement of God'. A first-hand account said: 'Immediately upon the cessation of the extremity of the earthquake, your heart would abhor to hear of the depredations, robberies and violences that were in an instant committed upon the place by the vilest and basest of the people; no man could call any thing his own, for they that were the strongest and most wicked seized what they pleased.'[3] For centuries, locals could see the ghostly ruins of Port Royal shimmering in the shallows. But it continued to sink deeper and deeper beneath layers of sand until it was completely lost from view, remaining as nothing more than a cautionary tale of how the wickedest city in the world met its doom.

But after Katrina, New Orleans was not lost, despite its supposedly sinful ways. When the floodwaters receded, they revealed a city that was down, but not out. Thick sludge coated the streets and houses. High-water marks stained the buildings. 'It was like a dirty ring on a bathtub all around the city,' Cassandra told me. She had been among the first to return to New Orleans in the aftermath of Katrina. The French Quarter was built on a natural levee and did not flood, so it was to this original settlement that returnees came and it was here that the first electric lights flickered on after the blackout. 'It was as if the city had gone back to the beginning and started all over again,' said Cassandra. Her description has the hallmarks of ancient flood legends where the survivors descend into a world cast into darkness, there to rekindle a flame, sow a seed and repopulate Earth.

After it returned to life, the French Quarter quickly resumed its status as the primary destination for party-loving tourists, much like Baia 2,000 years ago. Every day and every night of every week, revellers throng in a grid of Spanish colonial houses and Creole cottages, packed with music bars and eateries selling pizza slices, fried chicken or local sandwiches known as 'po' boys'. One night I fortified myself with a few cocktails and took a walk down the notorious Bourbon Street, crowded with people clutching plastic vessels shaped like hand-grenades. A Mexican Day of the Dead skeleton in a sombrero advertised 'jazz funeral daquiris'. Burly men with crewcuts guarded cavernous neon interiors thumping with house music. A stripper winked from the doorway of the Hustler Barely Legal club as I passed a man holding aloft a yellow python and a street performer humiliating some poor sap on a chair, cheered on by a hen party and bunch of lads in muscle T-shirts. There were live bands playing every twenty yards: rhythm-and-blues outfits, sixties-style girl groups in sequins, grunge cover bands, country rockers. On one corner, a street jazz band parped on trumpets, trombones and a tuba while dancing to a propulsive beat. Rivulets of beer ran along the asphalt where a nine-year-old kid on the kerb banged a drum, a hat full of loose change before him. A cheer went up as a woman hurled beads from a wrought-iron French balcony. Pungent clouds of skunk weed drifted from a parked car with speakers in the trunk blaring distorted hip hop while an uninterested cop sauntered by. It was a party at the end of the world behind walls that were braced against the might of a rising sea, where the future was of no importance, all bets were off, there were no more rules, and the music never stopped.

I crossed from the French Quarter into the Marigny district, where I settled myself with a beer at the only available table in a bar on Frenchman Street. On a dimly lit stage, a band played traditional New Orleans jazz. This music had originated when slaves from local sugar plantations gathered in the nearby Congo Square on Sundays, where they carried out voodoo ceremonies, played music from their origin countries and danced to exuberant polyrhythms. In the late nineteenth century, European jigs, reels and brass-band melodies were added to these African rhythms, birthing twentieth-century jazz. From that came its tributaries of rhythm and blues – pioneered by Professor Longhair and Fats Domino, a resident of the ill-fated Lower Ninth Ward – and the funky soul of Dr John and Allen Toussaint.

This city has given so much to the world, it is horrific to think that it might be on the brink of destruction. If New Orleans is to be saved, its defensive systems need to be brought in line with the realities of the topography and the changing climate. This will require immense funding and political will – two things that did not emerge following Hurricane Betsy or Hurricane Katrina.

If those disasters didn't inspire the appropriate action, then what could? The alternative is to depopulate and migrate. This is happening in Indonesia, where plans are afoot to move the rapidly sinking capital of Jakarta over a thousand kilometres north-east. 'If nothing is done, in the next 15 years, another 300 to 500 square miles of Louisiana will disappear,' wrote John M. Barry in the *New York Times*. 'And loss will continue after that, turning New Orleans into a potential Atlantis with walls of levees holding back the sea.'[4]

The band finished their set with a groovy, crowd-pleasing rendition of Herbie Hancock's 'Chameleon' while their trumpet player carried around the tip jar. When he returned to the stage at the end of their performance, he said, 'Thank y'all for coming. We'll be here again next week to bring you more music from below sea level.' But this was the sliver by the river, so we were technically above the sea, albeit by only a few metres. If the levees were to fail and the ocean rushed in, this crescent of higher ground would become an island – for a while, at least, until the Gulf of Mexico overwhelmed it. I imagined this band playing on, regardless, like the orchestra on the *Titanic*, as saltwater sloshed around the stage. Then further into the future, after the city was gone, when mariners will perhaps report the sound of old-time music on the waves, like the bells of Lyonesse and the Lowland Hundred, sounding a lament for a sunken city, and a warning for those yet to sink.

9

FUTURE FOSSILS

Humans had been nothing but trouble for God, ever since he set down the first man and woman in the Garden of Eden. When humanity became too wicked and corrupt for him to deal with any longer, God decided to unleash upon them a global flood and kill them all. But in his benevolence, he would spare the life of an innocent soul named Noah, along with his family.

God told Noah to build an ark made of gopher wood, and proportion it according to his specific instructions. Despite local ridicule, Noah constructed his ark. When it was finished, he gathered the male and female of every kind of living creature and took them into the boat with his wife and children to wait for the storm.

For forty days and forty nights, the rains came down and the flood came up. Afterwards, the land was entirely covered in water, even the highest mountains, and everyone apart from Noah's family was dead. When the waters receded, the ark came to rest on Mount Ararat.

After the flood, the world was repopulated again. But Noah's descendants did not trust God's plan for them to scatter across the earth. Nor did they believe God's promise that never again would he unleash such a deadly force against humanity. So they built an enormous tower that might unite them against their mercurial deity and protect them against a second flood.

Through the Tower of Babel, humankind would become god-like, using its powers of engineering to harness the natural world, and reshape reality in its own image.

But this only made God angry again.

Sixty thousand years ago, a forest grew where the Gulf of Mexico now laps the southern shore of Alabama. At that time, much of the planet's water was locked up in glaciers. The sea level was 120 metres lower, and the coastline was 100 kilometres further out. This was long before the oldest trees in the sunken forest of Pett Level were even seeded, when a dry, cold tundra connected England to France, roamed by mammoths and reindeer. It was slightly warmer in this part of North America. Tall cypresses dominated an inland swamp, teeming with life. Some of the trees had been standing for 500 years, perennial homes to squirrels and nesting egrets, aphids and mites, their trunks fluttering with bromeliads and orchids. But their days were numbered. A rapidly warming climate caused sea levels to rise. As saltwater inundated the swamp, the cypresses turned yellow and brown, then sickly and threadbare. Over eighty years the sea rose almost 2.5 metres, faster than the predicted worst-case scenarios of sea-level rise today. Then came the fatal blow. A major flood killed all the trees in a single year. They were buried in mud and sealed under sand as the sea pulled its watery blanket over their corpses.

The forest's existence would never have been known if it wasn't for Hurricane Ivan, which brought 27-metre waves crashing through the Gulf in 2004, churning up the seabed to expose over a square kilometre of stumps embedded in peat, 18 metres deep, perfectly preserved. Divers found trees with their bark intact, seeds and pollen, and traces of a riverbed. Pieces of cypress were raised to the surface. When researchers sawed the wood, it leaked sap and emitted the scent of resin. In that moment, what seemed a dead relic from a lost past became a visceral olfactory present as 60,000-year-old plant-communication systems suddenly reactivated, sending stress signals up the nostrils of twenty-first-century scientists. The trees contained thousands of years of history from the middle of the Ice Age in their growth rings, which could reconstruct past climate conditions and help us understand global warming in our own time. It's an illustration of how plants, animals and rocks can tell stories, too; tales of traumas and transformations experienced by non-human entities for millennia, and which will unfold long after humanity is but a ring of plastic in the fossil record. To consider the death of this forest, and others before it, is to understand how the planet has gone through many geological shifts and sea-level changes; cycles of evolution and extinction going back hundreds of millions of years to forests and wetlands that died, decomposed and turned into the fossil fuels that enabled the Industrial Revolution. A forest in the sea is therefore not only a glimpse of the distant past but one of the distant future, where decaying organic matter is in the process of becoming the rock, coal and oil of tomorrow.

Not far from that lost Alabama cypress forest, a similar phenomenon is taking place. In the neighbouring state of Louisiana, the cypress trees that once flourished in the wetlands of the Mississippi Delta are dying. Some 5,000 square kilometres of wetland south of New Orleans have been swallowed up by the Gulf of Mexico since the 1950s. An area of land the size of a football pitch disappears every 100 minutes. In part, this is down to a rise in sea level and an increase in the intensity of hurricanes. But the main cause is not Mother Nature, but the oil industry that has ravaged the wetland. This dying ecosystem is a shocking vision of ecological breakdown, where sea-level rise is not speculative but happening now, to human communities.

To steel myself for my journey to the furthest reaches of southern Louisiana, I found a channel on the car radio that specialised in New Orleans jazz, funk and soul. Trumpets blared and piano keys rolled as I headed south-west on Highway 90 to the city of Houma, then deep into the flatland. On both sides of the road were houses on stilts of concrete or wood, surrounded by featureless lawns with no plants or flowers, only fire hydrants and freestanding post-boxes amid the occasional live oak, casting a shadow on the gravel driveway. Beneath the larger homes were four-by-fours and tethered boats. Some properties were like static caravans, only metres above the ground, with junk and tools stored in their dark recesses. Much of the scenery was reminiscent of the East Anglian Fens: the long straight road, telegraph poles marching into an endlessly unfolding distance, an epic blue sky striated with cirrus streamers and floating islands of cumulus. Houses were accessed via wooden ramps over drainage ditches, where flood was a potential menace at the end of every driveway. Further south, there would be more water than land. Then no roads or houses at all. Only swathes of marsh grass and cypress trees in the brackish wetlands between Louisiana and the Gulf of Mexico, screeching with birds and clunking with oil derricks.

Closer to the edge of the landmass, the water channels grew wider. Egrets pondered the ripples from their verges. Occasionally, an angler on a jetty sat hunched over a rod. As the skyline flattened out into a ragged line of swamp grasses, the houses became fewer and more ramshackle. Some had been smashed to smithereens by high winds; wooden piles on the grass like funeral pyres. Many trees were stripped bare, skeletal against

the blue sky. Boulders lined the roadsides as a flood defence, making it hard to get a sense of the expanse of water, grasses and reeds around me. On digital maps this part of Louisiana looked like a decayed leaf, where the veins and venules remain, but the green lamina has disintegrated. When I studied satellite images, the wetland appeared as a murky blur, where land and sea were too uncertain to define, although I could make out clumps of vegetation pale beneath the surface, like recently drowned corpses.

In a best-case scenario, should the industrialised nations quickly cut carbon emissions and put in place all the recommended land-conservation measures, Louisiana would still lose most of its wetlands. Nobody can stop the subsidence or halt the rising sea. And, short of obliterating the levees that protect urban settlements to allow the Mississippi River to return to its natural cycles, it is not possible to resupply these wetlands with the sediments they need to become replenished. After a 2020 report into the effect of relative sea-level rise on the Louisiana wetland, Torbjörn Törnqvist, a Tulane University geology professor, said, 'We're screwed . . . the tipping point has already happened. We have exceeded the threshold from which there is basically no real way back any more, and there probably won't be a way back for a couple of thousand years.'[1] Gradually, but inevitably, the sea will take the land, swallowing small fishing communities until it reaches larger towns, and takes them too, just like the seas must have done in the English Channel, the Isles of Scilly and West Wales. In this transitioning terrain I could witness what happened thousands of years ago to many of the places in this book: forests poisoned by saltwater, flooded by tides, then submerged, along with human settlements and hundreds of years of cultural knowledge, leaving only folklore and fading memories. It's a glimpse of what the future might look like for low-elevation coastal areas around the world, in which 10 per cent of the population lives.

The most brutal effects of the sinking land could be seen in Isle de Jean Charles in the Terrebonne Parish, at the terminus of my route, which was on the verge of disappearing off the map entirely. A line of telegraph poles, like crucifixion crosses, led through the vanishing land. I could see immense tracts of open water with isles of bushy vegetation, bursting with clumps of flowering seaside goldenrod, streaked with barriers of swamp grass, waving in the wind.

Eventually, I turned onto Island Road, the only connection between Isle de Jean Charles and the rest of Louisiana. It was a 6-kilometre causeway, reinforced with verges of loose rocks, water splashing on both sides. I had seen footage on YouTube of waves inundating the tarmac and read stories of how this road could be completely submerged after storms and high tides. I had no idea if I could make it across to Isle de Jean Charles, or how much of the settlement I would find at the end of it. Indeed, as I approached the far end of Island Road, workmen were piling blocks of stone into high levees. It looked like a building site, not the entrance to a rural community. For a moment, I thought the way was barred, but they waved me past and down a single lane, glossy with an inch-high film of water. I wove around a few bends to find the first house, or what remained of it: a derelict frame of iron and wood, overgrown with foliage. Other homes were more intact, but obviously abandoned, with front doors swung open, glass broken, furniture and plastic children's toys strewn around the stilts. There were missing roofs. Collapsed balconies. Rusted struts. Holes torn in wooden side panels through which I could see the carcasses of kitchen units, dangling white power cords and peeling linoleum. A few houses had been reduced to piles of wood, as if a giant fist had come down from above and smashed them. But among the rubble, some homes were in good condition with cars parked outside. I thought I saw a shadowy figure move behind a net curtain, but I couldn't be sure. The only life I saw were two brown pelicans, which watched over a resting flock of ring-billed gulls at the end of a feather-littered jetty.

For almost 200 years this place had been home to descendants of the Biloxi, Chitimacha and Choctaw peoples, who fled their homelands after the Indian Removal Act in 1830. There followed two decades of government-instigated ethnic cleansing in which thousands died. Dispersed Native Americans joined up with the remnants of other Nations as they made their escape into the wilderness, seeking out seemingly inhospitable places to restart their lives beyond the murderous clutches of white colonists. One Native American, Pauline Verdin, married the son of a French immigrant named Jean Marie Naquin. They settled on a ridge of land in the marshes, which they named after

Jean's father, Jean Charles Naquin. When their children grew up, they also married Native Americans, creating a blended tribe in the wetlands, with French as their primary language. They planted corn, okra, beans and carrots. They grew herbs and made their own medicines. They harvested oysters and caught fish. They trapped muskrats and ducks. They farmed pigs and cows. They danced to Cajun music and gossiped at the local store. It became a thriving small community and by the 1910 census there were sixteen families and seventy-seven people living in the marsh.

It wasn't long before folklore took root. It was said that there was treasure buried on Isle de Jean Charles, belonging to the notorious privateer Jean Lafitte. Born in France in the late eighteenth century, he emigrated to New Orleans and embarked on a smuggling career. The Napoleonic Wars were raging and demand for British and French luxuries was high because of the 1807 Embargo Act, which forbade trade with both countries to protect the USA's commercial interests and avoid its ships being targeted. Lafitte quickly grew rich. Soon he owned ships, troops and weapons. But when the government cracked down on his trade in New Orleans, he re-established operations from an islet in the wetlands, close to what would become known as Isle de Jean Charles. In 1814, the US Navy intercepted Lafitte's fleet and captured many of his vessels. He escaped arrest and later was recruited by the government for the Battle of New Orleans, where he used his pirating skills to attack British ships. For a short time, he was an American hero but soon returned to his old ways. What happened next is uncertain, but it is believed that he – or one of his men – buried his treasure somewhere in this secluded spot in the wetlands. Some versions say that Lafitte murdered a rival and buried him with the treasure. The dead man's spirit protects the hoard. Should anyone attempt to dig for it, or even think about doing so, the haunted loot will sink deeper into the marsh. Today this fable reads like an allegorical warning about the effects of the oil and gas industry's digging in the wetland. Because if Lafitte's treasure ever did exist, it is likely that it is now beneath the sea, along with 98 per cent of Isle de Jean Charles.

In 1953, a road was built between the island and the mainland. But more destructive changes were afoot. Oil companies began to dredge canals through the marshes to transport their rigs, cutting through the isle, poisoning the freshwater with saltwater, killing trees and grasses, disrupting the delicate ecosystem. The people could no longer trap game or grow crops. Fracking, oil and gas extraction accelerated the subsidence of their land, shrinking their world, year after year. They enjoyed none of the oil wealth as compensation. Instead they were bullied into giving away access rights. The oil companies dug holes, drew out their prize, then abandoned them without filling in the canals, leaving conduits for saltwater and washing away what remained of the marsh. Local fisherman Jamie Dardar, born on Isle de Jean Charles, told the campaign group No Water No Life,

> If the oilfield companies had stuck to their agreement with people down here to fix their damage, and if they had done what they were supposed to do, we wouldn't be dealing with the loss of land on this island. But they didn't. They just came in; dug the canal; put up some wood to make a dam to stop the water from coming into their wells; and took off. All this land is eroding because of the oilfields.[2]

In the 1990s, the Army Corps of Engineers planned a system of levees to protect what remained of the wetlands. They didn't include Isle de Jean Charles, citing the high financial cost as their reason. Its fate was sealed. By 2018, it was reduced to less than a square kilometre. From a population of 350, the community numbered fewer than a dozen. Almost two centuries after they fled here, these Americans had lost their homeland again. They were the USA's first climate-change refugees; recipients of a federal grant to relocate 60 kilometres inland to Shriever, a former sugar-cane plantation. Louisiana State has rolled out similar plans for the Pecan Acres community in Pointe Coupee Parish, and Silver Leaf in Ascension Parish. These are among the first projects of many across the world that will become necessary when 200 million more people are displaced by global warming over the next three decades.*

There was so little left of Isle de Jean Charles that it didn't take long to reach the end of the road and turn back again. I saw no human beings. Only their traces. On the front of a dilapidated cabin, slightly tilted into the ditch behind it, was scrawled 'SOCIAL CLUB' and I could not tell if it was a bleak joke or a real venue. Set back in a nearby verge alongside a stream was a metallic orange pod with portholes like a 1960s Futuro house. This was a hurricane capsule, based on the design of a lunar module, designed to withstand high winds and high pressures. In the event of a tsunami, this metal ball would be carried through the flooded land, bobbing and rolling on the waves, until it came to rest like a tiny Noah's Ark, its lurid colour advertising its presence to any rescue boats and planes. Perhaps it would be the last thing to remain as the waters poured in and the road disappeared, floating like an SOS for a rescue that would be too late.

In a grassy expanse outside a derelict house behind a broken jetty was a sign from a busier era that read, 'No Parking – At Any Time'. The former residents of one abandoned home had

---

* Estimate by the United Nations.

left behind an artwork on their front lawn: a toilet on a wooden pallet plinth, wrapped in fairy lights that looked at first glance like barbed wire or a crown of thorns. A weathered placard above it read 'CLIMATE CHANGE NOT WORTH . . .'

I later discovered that was the work of Chris Burnet, who had reluctantly departed his family home for a new life in Shriever. The house had been built on the site of a garden that his grand-father used to tend before his parents got married. It was once full of trees that he climbed as a child. Those trees were all gone now. Killed by storms and floods. Some of his family's old pos-sessions remained in the space beneath the elevated house: an office chair, a car trailer, a Weber barbecue, a fridge, a cool box. Propped against one of the stilts was another handwritten sign:

ISLE DE JEAN CHARLES
IS NOT DEAD
CLIMATE
CHANGE
SUCKS

When I first decided to write about sunken lands, walking through a drowned Mesolithic forest near the site of a medieval coastal town, 16 kilometres from my home in England, I did not imagine that two years later I'd see with my own eyes a simi-lar phenomenon in current-day USA, where the structures of a hurricane-ravaged fishing town lay buckled in the mire and dead trees poked from pools of saltwater. This was a place where global-warming sea rise was no longer a potentiality, but a real event unfolding: saltwater flowing into the wetland, trees dry-ing, marshes sinking, refugees fleeing their homes, their culture surviving only in the memories of its scattered diaspora.

Despite this horror, my climate-change anxiety had not intensified. If anything, I felt more grounded. Instead of doom-scrolling social media feeds, allowing my imagination to catastrophise with wild abandon, global warming was visibly happening, right where I was standing, and I had no choice but to accept it. Rather than a paralysing terror, I felt a sense of resolve: an awareness of the preciousness of life, a determination to survive but also to make the best of what time was available. It is what Michael Dowd, a religious naturalist turned climate-grief advocate, describes as a 'post-doom mentality'. Doom, he says, is that understandable feeling of dread when we realise that our technological progress and economic development are causing climate breakdown and a mass extinction. But post-doom is what opens up 'when we remember who we are and how we got here, accept the inevitable, honour our grief, and prioritise what is pro-future and soul-nourishing'.

The story of Isle de Jean Charles is the story of what happened to so many people going back to the Last Glacial Maximum, when much of the world's water was locked up in walls of ice and humans settled in river valleys and coastal forests, harnessing the land for its food, medicinal plants and materials, populating it with their ancestral spirits. This might have been a scantly populated outpost in the wetlands but it was home to a unique blend of peoples who came here to escape persecution, who fished and farmed together, who held parties around boiling pots of shrimp and sang tribal songs, who knew everybody else's name, who told stories about haunted pirate treasure. What happened in this place mattered. It was a vision of the future that the world had to face up to with honesty, even if to do so was painful. 'You can't stop the water, so it's gotta come,' one Isle de Jean Charles resident told a filmmaker in 2013. 'You can't stop them storms. The water's gonna come in as high as it wants.'[3]

There are many others on planet Earth who know how she felt. The millions displaced in Nigeria after the worst floods for ten years. The thousands of south-eastern Australians who fled their homes after rivers broke their banks in New South Wales, Victoria and the island of Tasmania. The evacuees from El Castaño in Venezuela, whose town was washed away after heavy rains burst a nearby dam. The people in the coastal munici-pality of Da Nang, Vietnam, whose homes were submerged after storms. The residents of 450,000 homes in Thailand sunk by extreme monsoon rains and flash floods. The loved ones of those in Pakistan who died as floods wrecked villages, roads and bridges, ruining crops and severing supply routes, or the mil-lions of survivors in roadside tents facing thirst, malnutrition and disease from dirty drinking water. After that disaster, the UN Secretary General António Guterres said, 'We have waged war on nature and nature is striking back.'

This is the new reality. Since the Industrial Revolution we have released 2 trillion tonnes of carbon dioxide into the atmos-phere, adding the energy of 25 billion nuclear bombs to the biosphere in the last fifty years alone.[4] Almost all this energy has been absorbed by the oceans, sending temperatures soaring and sea levels rising, unleashing torrential rains and ferocious hur-ricanes. We are experiencing a transformation of Earth's climate that will transition our civilisation into something radically dif-ferent or bring about its collapse. This will transpire whether we believe we are responsible for global warming or not. We can argue about it all we like. The planet is getting hotter. Even if this process miraculously ceased overnight, our industrial activities would imperil us just the same.

In 2020, scientists from Chile and the UK used statistical modelling to predict the effect of our current rates of deforesta-tion and natural-resource consumption. They gave humanity a 'very low probability' of survival. 'Less than 10 per cent in the most optimistic estimate to survive without facing a catastrophic collapse,' they wrote. 'While the extent of human contribution to the greenhouse effect and temperature changes is still a matter of discussion, the deforestation is an undeniable fact.'[5] It is levee-building, dredging and oil extraction that has decimated the Louisiana wetlands and accelerated the destruction of Isle de Jean Charles, more so than rising seas, and it will be our compulsion for resource plunder that will, ultimately, doom us if it continues unabated. The irony is that towns and cities on flooding coastal wetlands, like Isle de Jean Charles, are those which are most likely to become preserved beneath the sea as the fossil fuels of the future.

—

Plaquemines Parish is a long, skinny limb of sub-delta in the Gulf of Mexico, dominated by the oil and gas industry. The Mississippi River runs through it like an artery for over 100 kilometres until it divides into multiple leveed channels, forming a bird's foot delta. This was the place sacrificed to save New Orleans during the 1927 Mississippi River flood, when engineers dynamited levees south of the city. It was where Katrina made landfall and where most of the population live below sea level while the wetlands are being eaten away around them. There are an estimated 5,000 oil wells in the region, and over 16,000 kilometres of canals scarring the marshes. Thanks to these intrusions, Plaquemines Parish is vanishing. Its residents live on a floodplain that is being eaten away on both sides and they are sitting ducks for storms. After Hurricane Ida struck in August 2021, the storm surge was so powerful that the Mississippi River reversed its flow. Water topped its banks, flooding the lowlands. High winds shattered houses and cut the electricity to medical centres with patients on ventilators, leading to frantic evacuations and a scramble for power generators. The communities of Ironton and Myrtle Grove were submerged. Coffins containing bodies of their loved ones were washed from their cement tombs and floated into their backyards. For a week, long stretches of Highway 23 remained under water. It was this road that would take me along the Mississippi River to Venice, a fishing community that shares its name with Europe's famous sinking city. From there, another road would take me to a spot on the map labelled 'Southernmost Point in Louisiana by Car', between a refinery and an abandoned oil wellhead. That was my final destination: literally, the end of the road.

That morning, heavy rain began to fall. Under a brooding sky, I barrelled down the highway, windscreen wipers dancing to the funky Allen Toussaint track on the local radio station. The highway was wide and flat, gently meandering as it tracked the river. I passed houses on stilts, mobile homes, and brick mansions on elevated earthworks. Nearby were shabby gas stations, fast-food joints, and the occasional church – blockish buildings of rendered concrete that seemed more like bunkers built to withstand the force of God's wrath than they were ecclesiastical erections to honour his benevolent grace. These settlements gave way to scrubland, overlooked by radio masts, electricity pylons and wooden poles strung with telephone wires, leading deep into the parish, where industrial facilities overshadowed the landscape.

In a swathe of marsh grass by the road, an oil-derrick pump bobbed up and down, like a robotic curlew, feeding on prehistoric liquified organic matter with its long proboscis. Distant ribbons of pale smoke heralded the chimneys and towers of oil refineries, with petroleum silos in rows behind barbed-wire fences. Enormous trucks pulled out of access roads, engines guttering. Pylons guarded electricity substations that looked like zoo enclosures for Daleks, the vertical prongs of insulators and lightning arrestors like death-ray plungers. Up above, jets of water from high-powered hoses formed arcs on the skyline, as if in a battle with the deluge.

A decade ago, Plaquemines Parish filed a lawsuit against Chevron USA, Shell, ExxonMobil Corp., ConocoPhillips Co. and BP America, demanding that these companies either fix the damage done to the wetlands, or compensate for their devastating losses.[6] This legal challenge is ongoing, and it is but one small battle in a bigger war. In 2023 US Attorney General Karl Racine filed a lawsuit against four major oil companies for systematically and intentionally misleading the public about how their products are a major cause of climate change. As far back as 1988, Shell's own report into greenhouse gases warned of 'significant changes in sea level, ocean currents, precipitation patterns, regional temperature and weather'. They saw clear evidence that the result would be the mass abandonment of coastal towns and cities, just as I'd seen the previous day in Isle de Jean Charles. Yet rather than investing in renewables, the oil and gas corporations have pushed for more fossil-fuel extraction, funding contrarian scientific reports that deny that global warming is caused by industrial activities. They have spent billions on advertising campaigns that either play down the links between their products and climate change or promote themselves as economically vital corporations working for a cleaner future. Despite the burgeoning number of lawsuits, Senate hearings and accusations of lies, propaganda and obfuscations, there is no sign that the oil and gas corporations' campaigns of disinformation and greenwashing will end, or that they will make any significant change to put the planet's health ahead of their profits. 'The industry has no real plans to clean up its act and is barrelling ahead with plans to pump more dirty fuels for decades to come,' says US Congresswoman Carolyn Maloney.[7]

Here I was in the heartland of that industry, where oil and gas facilities were being threatened by the elemental forces that they were helping to unleash. In 2021, the Phillips 66 refinery in Plaquemines Parish announced closure after the damage caused by Hurricane Ida, along with six other refineries in the Deep South. Everything around me, from the industrial structures and houses to the barrier islands and marshes, was on borrowed time.

By this point in my journey, the rain was beyond torrential, drilling into the car windows, blurring my view, blending the radio jazz with a monstrous percussive hiss. The road was coated in a film of dancing water. Effluent fountains of brown spray erupted from truck wheels. The light dimmed to a dismal murk. It was only 2 p.m. but all the vehicles' headlamps were on full beam, diffused by rain into frazzled orbs. Every now and then I'd hit a puddle and the car aquaplaned, forcing me to grip hard on the wheel. Just when I thought the deluge was peaking there was a roar as it came down with even more force, bouncing so hard off the asphalt that it appeared to be raining upwards. I slowed to a crawl, traversing the camber of the middle lane, where water could not gather – not yet anyway. But I worried that all might become awash and the engine would flood, bringing me to a standstill in the big wide nowhere.

Mile after mile, the rain kept falling as I pushed further into the southernmost regions where the land became fragmented and there was little to the east and west but river and sea. The signal from WWOZ FM crackled and faltered until it cut out completely. I couldn't see much through windows warped by rain, but there was a lot of open water to my right-hand side. To the left, between the long grass and corpses of salt-soaked oaks, I spotted flashes of concrete levee wall. Eventually, I reached a zone of trailer parks with a fire station and a dime store. This had to be Venice, but I couldn't tell for sure, because the satnav kept losing its connection. No matter. There was only one road and I was on it.

Suddenly, the levee wall swept me in a curve to where the highway terminated at a T-junction. A dock, lined with cranes, steamed in the rain. Boats bobbed on their moorings beside palm trees permanently bowed by the wind. I turned right, with my back to the Mississippi, and headed down a winding road through salvage yards and dry docks, with prefab buildings in gravel car parks littered with steel containers, black plastic bags and planks of wood. After a few kilometres, the land fell away and a narrow causeway led through brackish seawater, thick with grasses, towards the silos of a gas plant.

Not long ago, the view would have been of thick swampy forests. Now it was mostly sea and scattered cypress trees, poisoned by salt and petrochemicals. Many were leafless skeletons clustered with the silhouettes of cormorants. Some were stripped of branches, trunks sculpted into wooden tombstones by the waves. Others were mere nubs breaking the surface. It is no surprise that these are sometimes known as 'ghost forests'. The woodlands of the English Channel, the Isles of Scilly, Cardigan Bay and East Anglia might have looked much like this at one time, hanging onto life as floodwaters inundated their world. It felt as though I had come so far in space and time since my walk through the submerged English forest of Pett Level with my daughters, only to arrive at a scene that evoked its drowning many thousands of years ago.

The road left behind the open water for an expanse of marshland, where the steel sentinels of the Targa gas-processing plant towered above a blaze of floodlights. After I passed its access gate, the road deteriorated into a slush of mud and stones. Blurry in the downpour, a massive pile of gravel loomed ahead. I realised suddenly that this was the end of the road, but before I had

time to brake, I hit a massive body of water as high as the wheel arches, sending jets of filthy water up on both sides. I swung a hard left and pulled up onto a raised shingle bank, where the car came to a halt at a steep angle, dripping with water, motor spluttering. I recalled the many photographs I'd seen of stalled cars trapped on roads in storms as floodwaters swelled around them. Immediately, my dastardly imagination flash-forwarded the worst-case scenario: a freak wave would drown me where I sat. Saltwater would erode the car and tonnes of organic sludge would cover me. I'd lie for millions of years beneath silt and sea in a morass of decayed wetland mulch and decomposing sea creatures, slowly compressed into a dark layer of carbon-rich rock. Heat and pressure would break me down into hydrocarbons and transform me into oil. There I would lie in a slick of fossil fuel, ready to be plundered by some foolhardy future culture that has forgotten the mistakes of the past.

This masochistic montage took only a few seconds. Then I breathed deeply. The mundane truth was that I had driven through a large puddle and I needed to drive out again without flooding the engine, in which case I would stall and be forced to seek help in the nearest gas-extraction facility. Carefully, I manoeuvred the car in a tight three-point turn until the bumper was inches from the rippling brown expanse. The puddle seemed longer, wider and deeper than when I drove into it. This situation wasn't ideal, but I was in it now. I could not remain here while the rain kept falling and the water kept rising. I forced the gear into 'drive' and hit the accelerator.

Then I pushed forwards, regardless.

10

A DROWNED WORLD

On yet another sweltering morning, she pushed her wooden canoe into the lagoon and began to row across the sunken city. Most of it lay beneath the surface, where eels slithered through ruined tower blocks and crawfish scuttled over barnacled multi-storey car parks. But in the distance, a twisted metal spire, crawling with creeper vines, loomed like a forgotten ancient monument in an Amazonian jungle.

Two hundred years before the Great Floods, an Englishman named Ballard told a tale of what would happen in the future. His book The Drowned World described how a catastrophic solar event rapidly warmed the planet, melting all the ice and expanding the seas until all the cities were submerged.

Even though there were no longer any books, and no computers upon which to store information, the Englishman's prophesy lived on in the memory of those who endured the Great Floods. She heard the story from her mother, who heard it from her mother, who heard it from her mother, and so on and so forth, going back to that time when her ancestors lived in houses, drove cars and flew planes around the world.

She could not easily imagine that epoch as she drifted through the steaming mangroves in her canoe, clutching the oar with *translucent, webbed fingers. But sometimes when a skyscraper, radio mast or cathedral steeple broke through the surface, she could sense that vast, ancient city sprawled beneath her – glimpsed momentarily, like a dream – before it was lost again to the dark waters.*

As a little boy who wanted to become a writer when he grew up, my first story was about an underwater palace. When I was a man in my thirties I wrote my first published book about a marshland in London, littered with Ice Age artefacts and rural folklore that seeped through into the present metropolis. In my forties, I moved to the south coast, near a flooded Stone Age forest and realised that there was a mysterious drowned world all around the island I called home, full of stories that had been forgotten. As my children ran through the remains of that forest, I felt such fear for their future in a time of ecological breakdown that I decided to write one more story about sunken lands. I travelled from the cursed lakes and shores of Wales to the shrinking Fens of East Anglia to the sinking archipelago at the edge of the Atlantic. I swam around Roman ruins in the caldera of an active volcano near the entrance to hell and explored an American city on the edge of submergence in a flatland where people were already fleeing the rising sea.

On my journey I learned that our incessant drive to control the natural world, preventing its cycles of ebb and flow, had created an ecological disaster and placed millions of people at risk of suffering terrible floods. The dominant cultural and economic view of the biosphere as a material resource that could be plundered, broken down into parts, and reconstructed for human gain was a fundamental cause of the problem. The biosphere is an incomprehensibly complex living thing, like a god, and it could not be treated like this for long without terrible consequences.

In legends of sunken cities, the recurring message was that moral corruption and our refusal to live in harmony with nature was what angered the gods and condemned the people to earthquake, fire and flood. These were mythic emergency signals from the past, to which we needed to pay heed. Progress was not inevitable. Our place on the planet was neither divinely ordained nor certain. To visit a sunken ruin or walk through a submerged forest was to see that history, like nature, moves in cycles and nothing stays the same in a world defined by change. Humans had suffered terrible disasters in the deep past, caused by glacial meltwaters, seismic shifts and eruptions, when everything they knew had been lost. We can see their traumas repeating in the river valleys, wetlands and coastlines of the world today, where walls, dykes and ditches hold back the rising waters, while the imperilled people behind them either deny the threat or cross their fingers and hope for the best. To make matters worse, the industries that commit the most egregious resource plunder spend millions of dollars to suppress the truth about what is happening to the planet and pressure governments to maintain the deadly status quo rather than to consider radical changes that could help humanity survive.

What the world needs most are not new technologies that try to stave off the inevitable and feed our insane desire for continual industrial growth, but a fundamental change of mass consciousness, a shift in the way we view ourselves in relation to other species and other objects, so that we understand the intricate interconnectedness of things, reconsider our place in the biosphere and reassess what it means to be a good steward of the planet, if a steward is even required at all. One of the best ways to effect such a change is not through data that people don't understand and draconian laws that undermine their sense of freedom, but through storytelling, art, philosophy and culture. These have the power to make us see reality in new ways, gain empathy with other beings and other points of view, imagine alternative futures, and feel viscerally the horror of what is happening. They can motivate us to change our behaviour, organise our communities and join together in protest against those who cause us harm. But in nations such as the UK and the USA, the arts are being systematically undermined, underfunded and attacked by corporations, media barons and governments who wish to protect the system that most benefits them and keeps us fixated instead on phony culture wars, racist conspiracies and celebrity outrages that only enhance our sense of separation.

I started my journey with hope that I could contribute in a microscopic way to the groundswell of storytelling that might help us see what has gone wrong with our civilisation and understand what we might learn from our ancestors' experiences of global warming in the early Holocene. But as I travelled through sunken lands, I realised that I was far from the heroes I imagined in my childhood novels. The middle-aged me was a deeply anxious person, endemically pessimistic and prone to compulsive catastrophising. This made me either the ideal person to write this book or the worst. But perhaps my emotional confrontations with the realities of fallen civilisations and lost worlds would help others process their own feelings. Change is imminent and inevitable and there is no good in pretending otherwise. The Buddhist scholar Professor Rupert Gethin writes, 'As long as there is attachment to things that are unstable, unreliable, changing and impermanent, there will be suffering.' I had to stop clinging to that which could not remain the same, grieve honestly for what has been lost and then take part in a future in which I had agency: one that was a wide horizon, not an ever-narrowing channel of fate. But then my sunken-lands adventure ended in tragicomedy when I drove through the Louisiana wetland to the end of the world and fell over the edge.

Of course, I survived the ordeal of the big puddle to tell the tale. I returned home from New Orleans and set to work writing this book, ensconced in my study with a voodoo gris-gris bag for good luck. But if I thought the story was over, I was wrong. That year, the warmest ever recorded in the UK, ended with its coldest month in a decade, when an Arctic maritime airmass turned the streets to ice. Across the Atlantic, a bomb cyclone struck the USA and hurtled south at hurricane speeds towards the Mexican border. It was the deadliest winter storm in living memory. Blizzards and high winds cut power to over a million homes, trapping people in their cars and homes, killing multiple dozens. European temperatures soared to record highs, with the people of Poland and Switzerland experiencing a warmer January night than the average night in July. By that summer, the Greek island of Rhodes was on fire, the heaviest rain in known history had triggered floods and landslides in Japan, and the Hawaiian town of Lahaina had burned to the ground. A marine heatwave sent global oceanic temperatures soaring and swarms of jellyfish amassed on the south English coast where I lived. Energised by a summer of unprecedented temperatures, Storm Daniel struck Libya with force and the resulting floods killed thousands in the city of Derna, destroying buildings and sweeping bodies onto beaches almost a hundred kilometres away.

Weird weather like this will never stop because weirdness is how things are now. Climate change is not simply a case of rising temperatures, but freakish weather of all kinds, from sudden deep freezes and unseasonal droughts to extreme rainfall and violent storms. Multiple interconnected planetary systems are being thrown out of kilter. Shifting jet streams are disrupting the conveyor belts of cold and hot air around the globe. Increased salinity in the warmer sea has begun to alter the circulation of deep-ocean currents, further destabilising long-established climate patterns. Dry parts of the world will get drier and wet parts of the world will get wetter.

In the first weeks of the new year, as I tried to turn my copious travel notes into something that resembled a book, a downpour began and did not stop. Day after day, night after night, the rain came down and the floods came up. One night, the rain lashed so hard on my bedroom window that I struggled to sleep. After a slow morning of writing, I took advantage of a break in the deluge and headed to my local park to stretch my legs. I was astonished to find that it was under water. The tennis courts and five-a-side football pitch were submerged. Three separate duck ponds had become one lake, their central islands reduced to a few protruding weeping willow trees, their branches drooped in the swirling current. Streams had grown into raging rivers, frothing and spinning around the handrails of footbridges. The boating lake had overflowed its perimeter and spilled across the grass, sinking shrubs and litter bins. Stone steps, footpaths and walls glimmered below the scummy surface. Waves lapped the brass plaques of memorial benches on the path where I taught my daughters to ride their bikes.

All the streams that ran from this park flowed towards the sea through a culverted waterway beneath the town centre, then gushed out from a pipe in the shingle beach. That underground river had now burst up through the drains, submerging the Priory Meadow Shopping Centre. Before the Norman Conquest, the shopping centre's location had been the marshy edge of a bay. After the storms that sank Old Winchelsea in the thirteenth century, the bay was inundated with shingle to form a raised bank. In the Victorian era, it was paved, drained and filled with buildings.

Now the waters had returned to haunt this former wetland. An eel was filmed swimming near KFC and McDonald's, like a ghost from its aquatic past.

These scenes were evocative of *The Drowned World* by J. G. Ballard, his speculative novel in which global warming has flooded the planet and forced human survivors to flee to the poles. The climate has regressed to its Triassic state. Tropical jungles grow over sunken cities, populated by predatory iguanas and gigantic bats. The protagonist, Kerans, is on an ecological research expedition in a swamp where London used to be. The tops of its hotels, shopping centres and tower blocks loom from lagoons filled with alligators. Kerans is unmoved by the ruins because he was born after the flood and knows nothing else but the water world of his time. He has no wish to return to our urban past. Instead, he is determined to travel deeper into the tropics so that he might access the primal parts of the human brain from the earliest stages of our evolution. On a similar note,

M. John Harrison's 2020 novel *The Sunken Land Begins to Rise Again*, which inspired the title of this book, relates strange goings on in the docklands of London and the villages of Shropshire, where its ageing protagonists catch glimpses of fish-like humans entering the waterways, suggesting either an evolutionary leap forward, or a regression into our antediluvian origins.

Our earliest known ancestor was a microscopic creature known as *Saccorhytus* that lived 'between grains of sand'[1] on the seabed 540 million years ago. In 2014 palaeontologist Neil Shubin found the fossilised remains of *Tiktaalik*, a 375-million-year-old fish that contained all the base biological elements that would lead to the human form, including shoulders, elbows, legs, neck and wrist. 'Everything that we have are versions of things that are seen in fish,' says Shubin.[2] Even after our distant ancestors took to the land, water may have continued to shape us physical and mentally. There is a theory that humans took a different evolutionary pathway from our tree-dwelling ape cousins by adapting to watery coastal habitats, which is how we lost our body hair, attained our layer of subcutaneous fat and developed our unique nasal form.

If we hang on for thousands of years, after the seas have risen, who knows what new forms we might take. Perhaps we might experience reactivations of deeply embedded DNA from our aquatic origins in a flooded future in which we shall once again become water-dwelling animals. In South-east Asia, the Bajau people, referred to as 'sea nomads', have a remarkable ability to hold their breath for long periods under water so that they can dive for food. Studies have shown that genetic mutations are giving them larger spleens to store more oxygenated red blood cells. Another genetic change has boosted their mammalian diving reflex, where their heart rate slows and reserves of oxygen are sent to the brain and heart to keep them under water for longer. These traits will be passed down through future generations, when these specialised humans might eventually evolve into aquatic beasts like seals, whales or dolphins.

It is fascinating to speculate about what our world might look like far from now, after the climate has shifted to unimaginable extremes and the oceans have risen even further, inundating large cities and coastal towns, expanding lakes into inland seas, turning our parks, shopping centres, factories and residential streets into underwater ruins and intertidal mysteries. Who will we be? What will we remember? What might remain of our experiences in this epoch? What stories might survive? Perhaps it should be incumbent on artists living today to craft a mythos as a time-capsule warning to our descendants not to repeat our mistakes. This tactic is already being considered as a means of protecting future humans from nuclear facilities. In the 1980s, Hungarian linguist and semiotician Thomas A. Sebeok proposed formulating a cautionary folklore to deter people from meddling with toxic-waste storage sites. He reasoned that the most enduring organisations were religious institutions, and therefore an 'atomic priesthood' with its own legends and rituals could preserve the history and meaning of these sites. This sounds uncannily like the occult idea of Atlantis as an ancient, encoded warning message from a forgotten past, maintained by gnostic adepts in the Tibetan mountains. Perhaps we need to understand that message properly and do all we can to pay it forward with our own storytelling.

The weekend after my hometown flooded, the rains fell again, and my daughters grew restless. They were bored of me hiding in my study writing about submerged forests and ancient ruins. To entertain ourselves we decided to look through some old family photographs and videos, buried in a digital storage device I had not accessed for years. We huddled around my desktop computer, scrolling through photos of my ex-wife and me in our first flat, its lounge painted orange and white with psychedelic wallpaper. The girls laughed at the pudgy balls of flesh that were themselves as babies and cringed at videos of their toddler tantrums and sing-songs with toy guitars. There was a photo of my late grandmother, and another of my cocker spaniel as a puppy. I was surprised to see an old friend with whom I'd fallen out of touch, recalling how close we used to be. There were artefacts, too, that had once seemed so precious: a torpedo-shaped iPod dock, a SMEG fridge, 1950s-style American-diner furniture.

I realised that I was looking at images of a lost world. That marriage was over. Those babies were no longer babies. Our dog was dead. That decor had long ago been painted over by new owners. Those artefacts had been consigned to skips. The iPod was obsolete tech. That house was no longer my house. That city was no longer my home. Yet only now did I perceive that the world I once knew was no more. I never realised at the time because, in my lived experience, the end didn't come in the form of a single apocalyptic event. There was no Great Flood. No sudden wipe-out. The losses came in fits and starts. One after another. Gradual alterations of the present, barely discernible, that would produce enormous changes over time. And for every loss there was a gain. New friendships, interests, landscapes. New ways of being in the world. I had written books, recorded music, performed spoken word, travelled to weird and wonderful places, and enjoyed experiences that would never have been possible in the parameters of that old life. That lost past was not a dead husk but a chrysalis. Just as this life now was a transitional phase for whatever comes next.

The end of the world is not some dreadful threat on the hori-zon. It is already here. We live inside the unfolding catastrophe, which is why it is so hard to see. Day after day, the losses stack up. Since my second child was born, 160 species of animal have disappeared.[3] Thirty per cent of insects have gone.[4] Sixty-one million hectares of forest have been felled.[5] Fourteen trillion tonnes of ice have melted.[6] The ten warmest years in recorded history have occurred in the past ten years and the climate has measurably altered. The world is different from what it was a decade ago and in ten years' time it will be different again. In a hundred years it might be so radically transformed that whatever human culture exists on the planet will be unrecognisable.

But civilisations don't really die. They just change into something else. Every living person is an embodiment of all that has happened since our species evolved. We are an accumulation of experiences in an ongoing narrative. No matter how small we feel as individuals, we have the power to influence what happens next. Grains of sand we each might be, but together we are a beach. If we act together, our legacy need not be one of fire and flood, nuclear holocaust and mass extinctions, but of a harmonious culture that has learned from the mistakes of its past and worships the biosphere that sustains it with as much devotion as we might worship a god. This thought gives me a modicum of hope: a flicker of light in the dark, stormy sky, where a dove flaps its way towards higher ground.

SONGS FROM THE SUNKEN LANDS

Songs from the Sunken Lands is a dystopian disco-punk soundtrack to accompany the book. Mutant guitar riffs and pulsing chthonic beats will carry you on electronic currents through drenched landscapes, from the legendary Atlantis and the Roman town of Baia to the city of New Orleans, imperilled by global warming, via North American flood myths, biblical deluges and Mesolithic tsunamis.

Tracks:

1. The Great Serpent
2. Storegga Slide
3. Atlantis Again
4. Turn with the Tide
5. Nero's Nymphaeum
6. Aquatopia
7. Hurricane City
8. Children of the Flood

Written and performed by Gareth E. Rees. Additional vocals by Kirsty Hockenhull.

Songs from the Sunken Lands is available from Bandcamp. Visit the website or scan the QR code to listen for free:

https://garetherees.bandcamp.com/album/
songs-from-the-sunken-lands

ACKNOWLEDGEMENTS

Thanks to . . .

Simon Spanton and Ben Thompson for their words of encouragement.

Rafe Ward and Cassandra Snyder for the guided tours.

Pippa Crane, Sarah Rigby, Amy Greaves and the Elliott & Thompson team.

Kirsty Hockenhull, Pete and Penny Hockenhull, Mum and Dad, Paul Newland, Dave Allen, Rebecca Lambert, Dan Maudlin, Jane Campbell, Nick Laight and Harry Cockburn.

All the brilliant archaeologists who have worked on sunken places around the globe, making it possible for me to tell this story.

NOTES

Chapter 1: Children of the Flood

1. This story is based on a traditional Native American legend: https://www.native-languages.org/ojibwestory3.htm
2. 'Why the "Stonehenge of Atlit Yam" was Held in Such High Regard', Smithsonian Channel; https://www.youtube.com/watch?v=Dku5u7dF9Kw
3. Patrick Nunn, *The Edge of Memory: Ancient Stories, Oral Tradition and the Post-Glacial World* (Bloomsbury, 2018), page 136
4. 'Evidence Noah's biblical flood happened, says Robert Ballard', ABC News, December 2012; https://abcnews.go.com/Technology/evidence-suggests-biblical-great-flood-noahs-time-happened/story?id=17884533
5. 'Germany floods: "My city looks like a battlefield"', BBC website, 16 July 2021; https://www.bbc.co.uk/news/world-europe-57862570
6. Ciara Nugent, 'How deadly flooding in Germany and Belgium exposed Europe's climate change hubris', *Time* magazine, 19 July 2021; https://time.com/6081472/germany-flooding-climate-change

Chapter 2: The Forest Beneath the Sea

1. 'A Story of the Stone Age', published in Wells' collection, *Tales of Space and Time* (1920)
2. James Gallagher, '"Memories" pass between generations', *BBC News*, 1 December 2013; https://www.bbc.co.uk/news/health-25156510
3. 'Wooden Stone Age platform found on seabed off Isle of Wight', *BBC News*, 20 August 2019; https://www.bbc.co.uk/news/uk-england-hampshire-49405708
4. Clive Cookson, 'Evidence found of Stone Age wheat trade in UK', *Financial Times*, 26 February 2015; https://www.ft.com/content/ce67ab90-bd99-11e4-8cf3-00144feab7de
5. Brian Eno, 'The Big Here and Long Now'; https://longnow.org/essays/big-here-long-now
6. Damian Carrington, 'Rising seas threaten "mass exodus on a biblical scale", UN chief warns', *Guardian*, 14 February 2023; https://www.

theguardian.com/environment/2023/feb/14/rising-seas-threaten-mass-exodus-on-a-biblical-scale-un-chief-warns

Chapter 3: Lost Kingdoms

1. This is the author's own embellished version of the popular legend, combining different versions from multiple sources.
2. 'Brecknockshire', *Camden's Britannia* (1586); https://www.exclassics.com/camden/camden0082.htm
3. 'If Llangorse lake also shrunk during the dry period and expanded when wet conditions prevailed, human beings could have been there to see the changes.' F. J. North, *Sunken Cities* (University of Wales Press, 1956), page 112
4. J. Williams Ab Ithel (ed.), *The Barddas of Iolo Morganwg*, Vol. I, (1862); https://sacred-texts.com/neu/celt/bim1/bim1120.htm
5. 'Boddi Maes Gwyddno', *Black Book of Carmarthen*, trans. Jim Finnis; http://www.cantrer.pale.org/pages/poem
6. Patrick Nunn, *The Edge of Memory: Ancient Stories, Oral Tradition and the Post-Glacial World* (Bloomsbury, 2018), pp. 48–9
7. Felix Nobes, 'Borth and Ynyslas will be lost to the sea in decades, experts fear', *Cambrian News*, 13 November 2022: https://www.cambrian-news.co.uk/news/environment/borth-and-ynyslas-will-be-lost-to-the-sea-in-decades-experts-fear-572371
8. Lewis Morris, *Celtic Remains*, compiled 1717–57, transcribed by the author's nephew, 1778–9 (1878)
9. Jyoti Madhusoodanan, 'Top US scientist on melting glaciers: "I've gone from being an ecologist to a coroner"', *Guardian*, 21 July 2021; https://www.theguardian.com/environment/2021/jul/21/climate-crisis-glacier-diana-six-ecologist
10. John H. Richardson, 'When the end of human civilization is your day job', *Esquire* [August 2015], 20 July 2018; https://www.esquire.com/news-politics/a36228/ballad-of-the-sad-climatologists-0815
11. 'Madagascar is ground zero for climate injustice', *The Climate Reality Project*, 16 September 2021; https://www.climaterealityproject.org/blog/madagascar-ground-zero-climate-injustice
12. Hannah Moore with Thaslima Begum (presenters), 'Bangladesh's catastrophic flooding: the climate crisis frontline', *Guardian*, 6 July 2022; https://www.theguardian.com/news/audio/2022/jul/06/bangladeshs-catastrophic-flooding-the-climate-crisis-frontline
13. Richard Arlin Walker, 'Tribal nations adapt to being at "ground zero" of the climate crisis', *High Country News*, 14 April 2021; https://www.hcn.org/articles/climate-change-tribal-nations-adapt-to-being-at-ground-zero-of-the-climate-crisis
14. From the Welsh government's Fairbourne Coastal Risk Management Learning Project https://gov.wales/sites/default/files/publications/2019-12/fairbourne-coastal-risk-management-learning-project.pdf

15. Carey Baraka, 'A drowning world: Kenya's quiet slide underwater', *Guardian*, 17 March 2022; https://www.theguardian.com/world/2022/mar/17/kenya-quiet-slide-underwater-great-rift-valley-lakes-east-africa-flooding

16. Lisa O'Carroll, '"Nobody knows what happened": the row over the non-vanishing Irish lake', *Guardian*, 29 March 2022; https://www.theguardian.com/world/2022/mar/29/row-non-vanishing-irish-lake-lough-funshinagh-drain-flooding-environment-legal

17. David Owens, 'Family-of-three forced to evacuate their home in Carmarthen after flood destroys the ground floor', Wales Online, 14 October 2018; https://www.walesonline.co.uk/news/wales-news/family-three-forced-evacuate-home-15278903

18. Ian Lewis, 'Talks held over future of Carmarthen flood defences as experts warn "no easy solution"', *In Your Area*, 10 February 2021; https://www.inyourarea.co.uk/news/talks-held-over-future-of-carmarthen-flood-defences-as-experts-warn-no-easy-solution

19. Benjamin Franta, 'On its 100th birthday in 1959, Edward Teller warned the oil industry about global warming', *Guardian*, 1 January 2018; https://www.theguardian.com/environment/climate-consensus-97-per-cent/2018/jan/01/on-its-hundredth-birthday-in-1959-edward-teller-warned-the-oil-industry-about-global-warming

20. Geoffrey Supran and Naomi Oreskes, 'The forgotten oil ads that told us climate change was nothing', *Guardian*, 18 November 2021; https://www.theguardian.com/environment/2021/nov/18/the-forgotten-oil-ads-that-told-us-climate-change-was-nothing

21. Lisa Song et al., 'Exxon confirmed global warming consensus in 1982 with in-house climate models', *Inside Climate News*, 22 September 2015; https://insideclimatenews.org/news/22092015/exxon-confirmed-global-warming-consensus-in-1982-with-in-house-climate-models

Chapter 4: The Shrunken Fen

1. This story is based on Christopher Marlowe's *Legends of the Fenland People* (Cecil Palmer, 1926)

2. Ian D. Rotherham, *The Lost Fens* (The History Press, 2013), p. 8

3. The Fenland Black Oak Project; https://www.thefenlandblackoakproject.co.uk/black-oak

4. https://elfinspell.com/PrimarySourceDio.html

5. Jeremy Lent, *The Patterning Instinct* (Prometheus Books, 2017), p. 280

6. Ian D. Rotherham, *The Lost Fens* (The History Press, 2013), p. 22

7. Timothy Morton, *Dark Ecology* (Columbia University Press, 2016), p. 43

8. ibid., p. 39

9. Jim Hargan, 'The Fens: England Below Sea Level', *Historynet*, 4 May 2007; https://www.historynet.com/the-fens-england-below-sea-level

10. Stuart Anderson, 'Experts reveal Norfolk was devastated by mega-tsunami 8,000 years ago', *Eastern Daily Press*, 22 July 2020; https://www.edp24.co.uk/news/20752395.experts-reveal-norfolk-devastated-mega-tsunami-8-000-years-ago

11. Graham Swift, *Waterland* (Simon & Schuster, 1983), p. 157

12. James P. Graham, 'Losing Seahenge'; https://jamespgraham.com/project/losing-seahenge

13. John Clare, 'Winter in the Fens', in *John Clare: Major Works*, edited by Eric Robinson and David Powell (Oxford University Press, 2004)

Chapter 5: Atlantean Dreams

1. This is the author's own embellished version of the popular legend, combining different versions from multiple sources.

2. For more about this topic, read Akshay Ahuja's excellent essay 'Strange Children', published in *Walking On Lava: Selected Works For Uncivilised Times*, edited by Charlotte DuCann, Dougald Hine, Nick Hunt and Paul Kingsnorth (Chelsea Green Publishing, 2018)

3. John Michael Greer, *Atlantis: Ancient Legacy, Hidden Prophecy* (Llewellyn Publications, 2007)

4. Julian Jaynes, *The Origin of Consciousness in the Breakdown of the Bicameral Mind* (Mariner Books, 1976); https://www.julianjaynes.org/resources/books/ooc/en/the-causes-of-consciousness

5. Graham Readfearn, 'Melting Antarctic ice predicted to cause rapid slowdown of deep ocean current by 2050', *Guardian*, 30 March 2023; https://www.theguardian.com/world/2023/mar/30/melting-antarctic-ice-predicted-to-cause-rapid-slowdown-of-deep-ocean-current-by-2050

6. Bill Caraher, 'Sun Ra, Pseudoarchaeology, and Atlantis', Archaeology of the Mediterranean World; https://mediterraneanworld.wordpress.com/2021/09/06/music-monday-sun-ra-pseudoarchaeoogy-and-atlantis

Chapter 6: The Sinking Isles

1. This is the author's own embellished version of the popular legend, combining different versions from multiple sources.

2. Robert L. Barnett et al., 'Nonlinear landscape and cultural response to sea-level rise', *Science Advances*, vol. 6, issue 45, 4 November 2020; https://advances.sciencemag.org/content/6/45/eabb6376

3. ibid.

4. W. Booth, K. Adam, 'Sinking Tuvalu prompts the question: Are you still a country if you're underwater?', *Washington Post*, 9 November 2021; www.washingtonpost.com/world/2021/11/09/cop26-tuvalu-underwater

5. Pete McKenzie, 'A favourite reef, a beloved atoll: Marshall Islands parents name children after vanishing landmarks', *Guardian*, 25 March 2023; https://www.theguardian.com/world/2023/mar/25/a-favourite-reef-a-beloved-atoll-marshall-islands-parents-name-children-after-vanishing-landmarks

6. 'Adapt or die, says Environment Agency', Environment Agency, 13 October 2021; https://www.gov.uk/government/news/adapt-or-die-says-environment-agency

7. Dougald Hine, 'Remember the Future?'; https://dougald.nu/remember-the-future

8. Denise Chow, 'Odd "peeling" tectonic plate may explain Portugal's mysterious earthquakes', *NBC News*, 16 May 2019; https://www.nbcnews.com/mach/science/odd-peeling-tectonic-plate-may-explain-portugal-s-mysterious-earthquakes-ncna1006481

9. Jamie Bullen, 'Experts warn deadly tsunami could hit tourist hotspots in Spain and Portugal causing widespread devastation', *Mirror*, 28 March 2017; https://www.mirror.co.uk/news/uk-news/experts-warn-deadly-tsunami-could-10113197

10. Robert Lanza, 'How our memories hold the key to time', *Psychology Today*, 29 November 2021; https://www.psychologytoday.com/gb/blog/biocentrism/202111/how-our-memories-hold-the-key-time

Chapter 7: Inside the Volcano

1. The author's version of this story draws on Pliny the Younger's eyewitness account of the 79 CE Vesuvius eruption, incorporating descriptions from his text.

2. Dio 57.14.7–8 (Loeb translation)

3. Gregory S. Aldrete, *Floods of the Tiber in Ancient Rome* (Johns Hopkins University Press, 2007), p. 14

4. ibid.

5. 'Cornelius Tacitus: The Histories', Roman Britain; https://www.roman-britain.co.uk/classical-references/books-tacitus-histories/cornelius-tacitus-the-histories/

6. Gwyn Topham, 'More than 5,000 homes in England approved to be built in flood zones', *Guardian*, 22 November 2021; https://www.theguardian.com/society/2021/nov/22/more-than-5000-homes-in-england-approved-to-be-built-in-flood-zones

7. Anuradha Varanasi, 'Increasing numbers of US residents live in high-risk wildfire and flood zones. Why?', *News from Columbia Climate School*, 22 January 2021; https://news.climate.columbia.edu/2021/01/22/high-risk-wildfire-flood-zones

8. http://www.perseus.tufts.edu/hopper/text?doc=Perseus%3Atext%3A1999.04.0006%3Aentry%3Dbaiae

9. Sextus Propertius, 'Addressed to Cynthia', *Elegies*, trans. Vincent Katz; http://www.perseus.tufts.edu/hopper/text?doc=Perseus%3Atext%3A1999.02.0067%3Abook%3D1%3Apoem%3D11

10. Seneca, *The Tao of Seneca: Practical Letters from a Stoic Master*, vol. 1 (Loeb Classical Library, 1917); https://tim.blog/wp-content/uploads/2017/07/taoofseneca_vol1-1.pdf

11. Tara Cobham, 'Supervolcano which contributed to wiping out the Neanderthals has "realistic possibility" of erupting', *Independent*, 28 June 2023; https://www.independent.co.uk/news/science/volcano-campi-flegrei-italy-eruption-warning-b2365563.html

12. Jacqueline Ronson, 'Italy's supervolcano and the end of the Neanderthals', *Inverse*, 8 January 2017; https://www.inverse.com/article/26084-campi-flegrei-supervolcano-neanderthal-extinction

13. Kyle Harper, 'How climate change and plague helped bring down the Roman Empire', *Smithsonian* magazine, 19 December 2017; https://www.smithsonianmag.com/science-nature/how-climate-change-and-disease-helped-fall-rome-180967591

14. 'Could climate change have led to the fall of Rome?', *NPR*, 22 January 2011; https://www.npr.org/2011/01/22/133143758/could-climate-change-have-led-to-the-fall-of-rome

Chapter 8: Hurricane City

1. This story is based on a myth of the Choctaw: David I. Bushnell 'Myths of the Louisiana Choctaw', *American Anthropologist*, vol. 12, no. 4, 1910, p. 528

2. Edward Ward, *A Trip To Jamaica: With a True Character of the People and Island* (London, 1699)

3. Diana Paton and Matthew J. Smith (eds), *The Jamaica Reader: History, Culture, Politics* (Duke University Press Books, 2021)

4. John M. Barry, 'Is New Orleans safe?', *New York Times*, 1 August 2015

Chapter 9: Future Fossils

1. Mark Schleifstein, *Nola*, 22 May 2020: '"We're screwed": The only question is how quickly Louisiana wetlands will vanish, study says'; https://www.nola.com/news/environment/article_577f61aa-9c26-11ea-8800-0707002d333a.html

2. 'Isle de Jean Charles', *No Water No Life*, 13 September 2014; https://nowater-nolife.org/isle-de-jean-charles

3. *Can't Stop the Water* (Cottage Films, 2013); http://www.cantstopthewater.com

4. Karina von Schuckmann et al, 'Heat stored in the Earth system 1960–2020: where does the energy go?', *Earth System Science Data*, vol. 15, issue 4, 17 April 2023; https://essd.copernicus.org/articles/15/1675/2023

5. Jordan Davidson, 'Physicists: 90% chance of human society collapsing within decades', *EcoWatch*, 3 August 2020; https://www.ecowatch.com/human-society-collapse-deforestation-2646869167.html

6. Sibel Morrow, '5 major oil firms sued for causing climate change', *AA Energy*, 26 June 2020; https://www.aa.com.tr/en/energy/oil/5-major-oil-firms-sued-for-causing-climate-change/29705

7. Oliver Milman, 'Oil firms have internally dismissed swift climate action, House panel says', *Guardian*, 9 December 2022; https://www.theguardian.com/business/2022/dec/09/oil-gas-companies-fossil-fuel-industry-house-committee

Chapter 10: A Drowned World

1. 'Bag-like sea creature was humans' oldest known ancestor', University of Cambridge; https://www.cam.ac.uk/research/news/bag-like-sea-creature-was-humans-oldest-known-ancestor

2. Joe Palca, 'The human edge: finding our inner fish', *NPR*, 5 July 2010; https://www.npr.org/2010/07/05/127937070/the-human-edge-finding-our-inner-fish

3. Brian Resnick, 'The species the world lost this decade', *Vox*, 9 December 2019; https://www.vox.com/energy-and-environment/2019/12/9/20993619/biodiversity-crisis-extinction

4. Damian Carrington, 'Plummeting insect numbers "threaten collapse of nature"', *Guardian*, 10 February 2019; https://www.theguardian.com/environment/2019/feb/10/plummeting-insect-numbers-threaten-collapse-of-nature

5. Hannah Ritchie and Max Roser, 'Deforestation and forest loss'; https://ourworldindata.org/deforestation

6. Chris Mooney and Andrew Freedman, 'Earth is now losing 12 trillion tons of ice each year. And it's going to get worse', *Washington Post*, 25 January 2021; https://www.washingtonpost.com/climate-environment/2021/01/25/ice-melt-quickens-greenland-glaciers

SELECT BIBLIOGRAPHY

I read a lot of books while I was researching and writing *Sunken Lands*. Not all were directly referenced in the text, but they informed my thoughts and ideas. Below is a list that includes novels, poetry, non-fiction, websites, podcasts, films and albums that you might like if you enjoyed this book.

Books

Aldrete, Gregory S., *Floods of the Tiber in Ancient Rome* (Johns Hopkins University Press, 2007)

Ballard, J. G., *The Drowned World* (HarperCollins, 1962)

Baxter, Stephen, *Flood* (ROC, 2009)

Bonnett, Alastair, *The Age of Islands* (Atlantic Books, 2020)

Blavatsky, H. P., *The Secret Doctrine*, abridged by Michael Gomes (Penguin, 2009)

Boyce, James, *Imperial Mud: The Fight for the Fens* (Icon Books, 2020)

Cayce, Edgar, *Atlantis* (A.R.E. Press, 2009)

Clare, John, *John Clare: Poems*, selected by Paul Farley (Faber, 2007)

Cox, Edward W., 'Traces of submerged lands on the coasts of Lancashire, Cheshire and North Wales', *Transactions of the Historic Society of Lancashire and Cheshire*, vol. 46 (1894), pp. 19–56

Crary, Jonathan, *Scorched Earth: Beyond the Digital Age to a Post-Capitalist World* (Verso, 2022)

Crowley, Aleister, *The Drug and Other Stories* (Wordsworth Editions, 2015)

Davies, Edward, *The Mythology and Rites of the British Druids* (J. Booth, 1809)

Eggers, Dave, *Zeitoun* (Penguin, 2009)

Farrier, David, *Footprints: In Search of Future Fossils* (4th Estate, 2020)

Flynn, Cal, *Islands of Abandonment* (HarperCollins, 2021)

Green, Matthew, *Shadowlands: A Journey Through Lost Britain* (Faber & Faber, 2022)

Greer, John Michael, *Atlantis: Ancient Legacy, Hidden Prophecy* (Llewellyn Publications, 2007)

Halperin, Ilana, *Felt Events* (Strange Attractor, 2022)

Hancock, Graham, *Underworld* (Penguin, 2002)

——, *Magicians of the Gods* (Coronet, 2015)

Harman, Graham, *Towards Speculative Realism* (Zero Books, 2009)

Harrison, M. John, *The Sunken Land Begins to Rise Again* (Gollancz, 2020)

Horowitz, Andy, *Katrina: A History 1915–2015* (Harvard University Press, 2020)

Itäranta, Emmi, *The City of Woven Streets* (Harper Voyager, 2016)

Krznaric, Roman, *The Good Ancestor: How to Think Long Term in a Short-Term World* (Penguin, 2020)

Lent, Jeremy, *The Patterning Instinct* (Prometheus Books, 2017)

Lovecraft, H. P., *Necronomicon: The Best Weird Tales of H. P. Lovecraft* (Gollancz, 2008)

Marcus Aurelius, *Meditations* (Penguin, 2006)

Marlowe, Christopher, *Legends of the Fenland People* (Cecil Palmer, 1926)

McGilchrist, Iain, *The Master and His Emissary: The Divided Brain and the Making of the Western World* (Yale University Press, 2019)

Morton, Timothy, *Dark Ecology* (Columbia University Press, 2016)

——, *Humankind* (Verso, 2017)

North, F. J., *Sunken Cities* (University of Wales Press, 1957)

Nunn, Patrick, *The Edge of Memory: Ancient Stories, Oral Tradition and the Post-Glacial World* (Bloomsbury, 2018)

Read, Rupert, and Samuel Alexander, *This Civilisation is Finished* (Simplicity Institute, 2019)

Reid, Clement, *Submerged Forests* (Cambridge University Press, 1913)

Rhys, John, *Celtic Folklore: Welsh and Manx* (Clarendon Press, 1901)

Rotherham, Ian D., *The Lost Fens* (The History Press, 2013)

Solnit, Rebecca, *Hope in the Dark* (Canongate, 2005)

——, and Rebecca Snedeker, *Unfathomable City: A New Orleans Atlas* (University of California Press, 2013)

Swift, Graham, *Waterland* (Simon & Schuster, 1983)

Teilhard de Chardin, Pierre, *The Phenomenon of Man* (HarperCollins, 1976)

Trevelyan, Marie, *Folk-Lore and Folk-Stories of Wales* (E. Stock, 1909)

Trigg, Dylan, *The Memory of Place* (Ohio University Press, 2012)

van Lohuizen, Kadir, *After Us the Deluge* (Lannoo, 2021)

Virgil, *The Aeneid*, rev. edn (Penguin Classics, 2003)

Online

'The Dark Mountain Project', dark-mountain.net

Future Fossils podcast, https://michaelgarfield.substack.com/podcast

The Long Now Foundation, longnow.org

Losing Seahenge (short film by James P. Graham), jamespgraham.com/project/losing-seahenge

'Native Languages of the Americas: Preserving and Promoting American Indian Languages', native-languages.org

Post Doom: Regenerative Conversations Exploring Overshoot Grief, Grounding, and Gratitude, postdoom.com

Weird Studies podcast, weirdstudies.com – a fount of inspiration and ideas.

Films

20,000 Leagues Under the Sea, dir. Richard Fleischer (1954)

At the Earth's Core, dir. Kevin Connor (1976)

Can't Stop the Water, dir. Rebecca Marshall Ferris, Jason Ferris and Kathleen Ledet (2013), cantstopthewater.com

The Fabulous Journey to the Centre of the Earth, dir. Juan Piquer Simón (1977)

Isle de Jean Charles, dir. Emmanuel Vaughan-Lee (2014)

The Land That Time Forgot, dir. Kevin Connor (1975)

Albums

Drexciya, *The Quest* (Submerge, 1997)

Gwenno, *Le Kov* (Heavenly Recordings, 2018)

Hawkwind, *In Search of Space* (United Artists Records, 1971)

Kemper Norton, *Toll* (Front and Follow, 2016)

The Lowland Hundred, *Under Cambrian Sky* (Victory Garden Records, 2010)

——, *The Lowland Hundred* (Exotic Pylon, 2014)

Sun Ra & His Astro Infinity Arkestra, *Atlantis* (Saturn Research, 1969)

Sunken Lands Playlist

Search for 'Sunken Lands' on Spotify to find my playlist of music to accompany this book:

1   Bessie Smith, 'Back Water Blues'
2   The Golden Gate Quartet, 'Noah'
3   Sun Ra, 'Lemuria'
4   Donovan, 'Atlantis'
5   Jimi Hendrix, '1983 . . . (A Merman I Should Turn to Be)'
6   Hawkwind, 'We Took the Wrong Step Years Ago'
7   Creedence Clearwater Revival, 'Born on the Bayou'
8   Nick Cave & The Bad Seeds, 'Muddy Water'
9   Jon Hassell, Brian Eno, 'Delta Rain Dream'
10   Kemper Norton, 'The Town'
11   Gwenno, 'Hi a Skoellyas LIV a Dhagrow'
12   David Bowie, 'Ashes to Ashes'
13   Parliament, 'Aqua Boogie (A Psychoalphadiscobetabioaquadoloop)'
14   Ghostface Killah, 'Underwater'
15   Drexciya, 'The Last Transmission'
16   LTJ Bukem, 'Atlantis (I Need You)'
17   Smallhaus, 'These Sunken Lands'
18   The Lowland Hundred, 'Anemone'

INDEX

INDEX